Practice Scenarios

I0049949

Dedication

There is only one reason why a firefighter would purchase this book. They are serious about improving their ability to tackle and overcome challenges. Because of that, I dedicate this book to you. My goal is to help you achieve your professional goals and I sincerely hope this book helps you do just that.

About the book

Practice Scenarios was developed to help prepare dedicated firefighters of all ranks to handle problems and challenges in four major areas: Fire Incidents, Non-Fire Incidents, Supervision and management, and Administrative tasks.

This is meant to be a workbook for you to practice skills. It is not meant to take the place of an educational fire service text. Practice Scenarios is designed to help you take the knowledge you already have and package it into workable formats that can be used in the field and during promotional interviews and examinations. Each scenario provides a challenging incident, followed by an answer key. After each answer key are fireproof tips that are designed to provide you with an edge over your competition and help elevate your situational awareness on the fire ground.

When it comes time for you to take your promotional exam, preparing to present your answer is vitally important. You don't want to make the mistake that so many others do, which is, showing up on test day unprepared. If the first time you give an oral presentation is on the day of your exam, you are not playing your hand well. This book will help prepare you to hit the main points that will help you make a great impression and provide the best possible score. This is accomplished by not only knowing your job, but also delivering your message in the most professional manner possible.

If you are familiar with my work, you know I am a fan of simplification. This is why I use acronyms and formats to help me organize my thoughts and action plan. You will notice that some of the acronyms I use in this workbook are different than those in my tactical book Fireground Operational Guides. This should serve as a reminder that there is more than one way to reach your destination. Although the acronyms differ, the end results are the same. All the key areas that you must address are covered in both books.

I had been in the fire service for more than twenty years when I developed this book. The evening before writing this introduction, my crew and I responded to a chemical spill in a shipping warehouse. I utilized the hazardous material scenario format outlined in this book to carry me through that incident. It's the same format I used to attain a perfect score in my Deputy Chief promotional exam. In other words, many of these formats work in the field as well as they do in the assessment center. I sincerely hope you find these scenarios and formats as helpful as I do.

Disclaimer

The recommendations, advice, descriptions, and the methods in this book are presented solely for educational purposes. The intent of this text is to help the reader organize the way they approach and prepare for promotional interviews. This is NOT a tactical book and should be treated as such. The reader is responsible to follow the rules, regulations, policy's and operating guidelines of the organization they belong to. The author and publisher assume no liability whatsoever for any loss or damage that results from the use of any of the material in this book. Use of the material in this book is solely at the risk of the user.

Disclaimer

The recommendations, advice, descriptions, and the methods in this book are presented solely for educational purposes. The intent of this text is to help the reader organize the way they approach and prepare for promotional interviews. This is NOT a tactical book and should be treated as such. The reader is responsible to follow the rules, regulations, policy's and operating guidelines of the organization they belong to. The author and publisher assume no liability whatsoever for any loss or damage that results from the use of any of the material in this book. Use of the material in this book is solely at the risk of the user.

Scenario Index

Assignment Card

For all scenarios, it should be assumed that the department you work for operates with 4 three-member Engine companies and 2 four-member ladder companies. We will also use the designation of Deputy Chief (DC) and Battalion Chiefs (BC). The DC will also be considered the tour commander.

In each scenario, YOU will fulfill the role of Incident Commander (IC). This will help you prepare for scenarios by helping you see and understand the big picture. In some incidents, you will be assigned as the DC or BC. In other incidents, you may simply be designated as the tour commander. Don't ever forget that an officer at any rank may be asked to assume the role of the IC at any given incident.

1st Alarm	E-1, E-2, L-1, DC-1, BC-1
2nd Alarm	E-3, E-4, L-2, BC-2
3rd Alarm	E-5, E-6, L-3, BC-3 (Mutual Aid Companies)
4th Alarm	E-7, E-8, L-4, BC-4 (Mutual Aid Companies)

Note: Additional resources that may be needed, such as a rescue company or haz-mat team, will be considered mutual aid companies and must be requested by the IC.

Fire Scenarios

The Format

In our best-selling book *Fireground Operational Guides*, retired Deputy Chief Mike Terpak and I introduced the reader to 17 areas that must be addressed at every structure fire in order to ensure success. Those same 17 categories must also be addressed when tackling fire scenarios. For simplicity, I condensed some of those categories together and developed a simple format to help you prepare for various fire scenarios. Creating and memorizing bulleted formats and acronyms have always helped me establish a rhythm that has helped me both in the field and when fighting the "paper fire" in front of a promotional panel. Although your strategy and tactics will change at every incident, the format you follow can remain the same. This will also help ensure consistency that will enable you to memorize some of the constants at each fire (like the resources you call) and focus on the specifics of each individual scenario.

Below is the list of 10 areas you must address at every structure fire, followed by some clarifying thoughts and some scenarios to help practice. You don't have to use this sample format, but I strongly encourage you to develop one that works for you.

1. **<u>En route</u>** – RPM

   There are a few things you should do on your way to the fire scene: Monitor <u>R</u>adio Reports, Review <u>P</u>re-incident plans of the building/area, and attempt a <u>M</u>ulti-sided view of the structure.

2. **<u>Upon Arrival</u>** – Position IC CAR PPE

   Establish Command if you are the first to arrive. If you are relieving a current Incident Commander (<u>IC</u>), have a brief face-to-face meeting with that individual for a quick incident briefing where the following information is relayed: Conditions upon arrival, Actions already taken, Resources already requested (<u>CAR</u>). In other words, find out answers to the following three questions: What do you have? What did you do? *and* What do you need? When relieving another IC, be sure to *Assume* Command and assign that individual to another position (such as operations or interior division chief). You should also be conscious of where you <u>Position</u> your vehicle to ensure it allows for the proper positioning of other apparatus on the scene. Also ensure that all members who work on scene wear appropriate <u>PPE</u> for the duration of the incident.

3. **Initial Radio Report** – FM Radio

Establish your tactical radio Frequency. Determine your operational Mode (Offensive, Defensive, Combination, or Non-intervention). Then begin transmitting your Initial Radio Report (IRR), which should consist of the following information: floors and construction type, fire conditions, obvious size up concerns, and the name/location of the command post.

Note: Also remember to request periodic progress reports throughout the incident.

4. **Size-Up** – COAL TWAS WEALTHS

Conduct a thorough 15 point size-up and address any conditions that require your attention. The 15 points are: Construction, Occupancy, Apparatus/staffing, Life hazard, Terrain, Water supply, Auxiliary appliances, Street conditions, Weather, Exposures, Area, Location/extent of fire, Time of day, Height and Special considerations.

5. **Resources** – A RULES WAR+

There are resources you should call for at every working fire, regardless of size and scope. The acronym above is one way to remember them. These are not the only resources you will call (or assign) at a working structure fire, but rather the minimum list of the essential ones.

- ☐ Additional alarms - to ensure an adequate number of personnel and resources.

- ☐ Rapid Intervention Crew (RIC) - for firefighter safety.

- ☐ Utility Companies (gas, electric, water) - for utility control.

- ☐ Law enforcement - for traffic/crowd control.

- ☐ EMS (BLS/ambulance, ALS/paramedic) - for patient triage and treatment.

- ☐ Safety Officer (SO) - for scene safety.

- ☐ Water Supply Officer (WSO) - to ensure adequate water.

- ☐ Accountability Officer - to track and account for all on scene personnel.

- ☐ Rehab Unit – for rest and rehabilitation.

The plus (+) represents all additional resources that will be specific to the scenario. Some examples include: Red Cross (for displaced occupants), Dept. Public Works (for sand and salt), Coast Guard, Haz Mat Team, Dept. of Environmental Protection, USAR, Rescue Co, etc.

6. **Strategy / Tactics**

The best incident management system (IMS) in the world is worthless without sound objectives, strategy and tactics. This is where a well prepared and intelligent officer will shine. Begin by listing the objectives you are trying to accomplish (highest priority first). This should be followed by specific company assignments and tactics.

Examples of objectives:

- ☐ Remove civilians from danger.

- ☐ Protect the stairwell.

- ☐ Locate/Contain the fire.

- ☐ Ventilation the building.

Engine Company Ops: The main goals of all engine companies at most incidents will be to secure a water supply; locate, confine and extinguish the fire; and protect life and exposures. Here is a list of the key points you will want to address with engine companies.

- ❑ Position apparatus.

- ❑ Establish a primary (and secondary) water supply.

- ❑ Initiate attack.

- ❑ Choose appropriate size hose lines.

- ❑ Advance and position hose lines .

- ❑ Locate, Confine, & Extinguish (LCE) the fire.

- ❑ Protect life hazards.

- ❑ Protect exposures.

- ❑ Supply auxiliary appliances.

- ❑ Utilize Thermal Imaging Cameras (TIC).

- ❑ Coordinate with ladder companies (and other companies on scene).

- ❑ Provide periodic progress reports.

Ladder Company Ops: The goals of all ladder companies at most incidents can be remembered by using the acronym LOVERS-UPS (TIC-COP). Note: That acronym, underlined below, is not in order.

- ❑ <u>P</u>osition apparatus.

- ❑ Raise and position <u>L</u>adders.

- ❑ Force <u>E</u>ntry.

- ❑ Primary and Secondary <u>S</u>earch.

- ❑ <u>R</u>escue Operations.

- ❑ <u>V</u>entilation (Horizontal, Vertical, Outside Vent).

- ❑ <u>U</u>tility control.

- ❑ <u>S</u>alvage.

- ❑ <u>O</u>verhaul.

- ❑ Utilize <u>TIC's.</u>

- ❑ <u>C</u>oordinate with engine companies (and other companies on scene).

- ❑ Provide periodic <u>P</u>rogress reports.

7. **Incident Management** – U SIL FLOP SR

Use the Incident Command System at ALL structure fires, regardless of size/scope.

Assignments may include the following command and support staff members:

- ❑ <u>U</u>nified Command Staff (should be considered).

- ❑ <u>S</u>afety officer (to ensure a safe working environment).

- ❑ <u>I</u>nformation officer (to communicate with the media and concerned citizens).

- ❑ <u>L</u>iaison officer (to communicate with other agencies and ensure smooth operations).

- ❑ <u>F</u>inance officer/section.

- ❑ <u>L</u>ogistics officer/section.

- ❑ <u>O</u>perations officer.

- ❑ <u>P</u>lanning officer/section.

❑ Staging officer.

❑ Rehabilitation officer.

Also: Establish Divisions/Groups early to enhance communications and improve accountability. Don't wait to be overwhelmed with your span of control. (Example: Rescue = Rescue Group; Vertical ventilation = Roof Division, etc.) Assigning division supervisors will help you manage your span of control. Less people trying to communicate with you directly ensures a greater level of effectiveness.

8. **Benchmarks** – Time, Progress, RER/LIP

Time management helps the IC measure if tactics are working. Review, Evaluate, and Revise your strategy and tactics periodically and use progress reports to ensure you are meeting your objectives of Life safety, Incident Stabilization, and Property Conservation (RER/LIP).

❑ <u>Time</u> management: Note time of attack. Request <u>Progress</u> reports every 10-15 min.

❑ <u>RER/LIP</u>

9. **Under Control** – A SOS PAR ICS DO

When the fire is under control, <u>A</u>nnounce via radio then conduct the following:

❑ Complete <u>S</u>econdary Searches.

❑ Meet <u>S</u>alvage and <u>O</u>verhaul responsibilities.

❑ Conduct a Personnel Accountability Roll Call (<u>PAR</u>) to account for all personnel.

❑ Request an <u>I</u>nvestigation Unit for cause and determination.

❑ Check <u>C</u>O levels.

❑ <u>S</u>ecure the Building.

❑ <u>D</u>emobilize the Incident.

❑ Turn the building over to the <u>O</u>wner.

10. **Terminating/Transferring Command** – TD DOP

- ❏ **T**ransfer to another officer or Terminate completely.
- ❏ Conduct an incident **D**ebriefing.
- ❏ **D**ocument the incident, complete reports.
- ❏ **O**ffer Critical Incident Stress Debriefing (CISD).
- ❏ Schedule a **P**ost Incident Analysis (PIA)

The Format – Short Version

Once you understand the format above, you will want to condense each of the 10 categories down to the least amount of words possible. This will enable you to create a tactical worksheet for real world incidents, as well as practice and promotional scenarios. Here is an example of what it might look like.

1. **En route** – RPM

- ❏ Monitor **R**adio Reports
- ❏ Review **P**re-incident plans
- ❏ **M**ulti sided view

2. **Upon Arrival** – Position IC CAR PPE

- ❏ **Position** vehicle properly
- ❏ Establish Command (IC), or
- ❏ Face-to-face with current IC
- ❏ Incident Debriefing - Conditions, Actions, Resources (CAR)
- ❏ Assume Command
- ❏ Re-assign the IC
- ❏ Ensure all members wear PPE and respiratory protection

3. **Initial Radio Report** – FM Radio

   ☐ Establish Tactical Radio <u>F</u>requency

   ☐ Determine/Announce Operational <u>M</u>ode: (O/D/C/N)

   ☐ IRR: Floors and construction type, Fire conditions, size up concerns, name and Location of the CP

4. **Size-Up** – COAL TWAS WEALTHS

   ☐ <u>C</u>onstruction, <u>O</u>ccupancy, <u>A</u>pparatus/staffing, <u>L</u>ife hazard, <u>T</u>errain, <u>W</u>ater supply, <u>A</u>uxiliary appliances, <u>S</u>treet conditions, <u>W</u>eather, <u>E</u>xposures, <u>A</u>rea, <u>L</u>ocation and extent of fire, <u>T</u>ime of day, <u>H</u>eight and <u>S</u>pecial considerations.

5. **Resources** – A RULES WAR+

   ☐ <u>A</u>dditional alarms - to ensure adequate personnel.

   ☐ <u>R</u>apid Intervention Crew (RIC) - for firefighter safety.

   ☐ <u>U</u>tility Companies (gas, electric, water) - for utility control.

   ☐ <u>L</u>aw enforcement - for traffic/crowd control.

   ☐ <u>E</u>MS (BLS/ambulance, ALS/paramedic) - for patient triage and treatment.

   ☐ <u>S</u>afety Officer (SO) - for scene safety.

   ☐ <u>W</u>ater Supply Officer (WSO) - to ensure adequate water.

   ☐ <u>A</u>ccountability Officer - to track and account for all on scene personnel.

   ☐ <u>R</u>ehab Unit – for rest and rehabilitation.

   ☐ (±) Red Cross (for displaced occupants), Dept. Public Works (for sand and salt), Coast Guard, Haz Mat Team, DEP, USAR, Rescue Co, etc.

6. **Strategy / Tactics** - List the objectives and assign companies

   Primary objectives include:

   ☐ (scenario dependent)

Engine Company Ops:

- [ ] Position apparatus
- [ ] Establish a primary (and secondary) water supply
- [ ] Initiate attack
- [ ] Choose appropriate size hose lines
- [ ] Advance and position hose lines
- [ ] Locate, Confine, & Extinguish the fire
- [ ] Protect life hazards
- [ ] Protect exposures
- [ ] Supply auxiliary appliances
- [ ] Utilize TIC's
- [ ] Coordinate with ladder companies (and other companies on scene)
- [ ] Provide periodic progress report

Ladder Company Ops: LOVERS-UPS (TIC-COP)

- [ ] <u>P</u>osition apparatus
- [ ] Raise and position <u>L</u>adders
- [ ] Force <u>E</u>ntry
- [ ] Primary and Secondary <u>S</u>earch
- [ ] <u>R</u>escue Operations
- [ ] <u>V</u>entilation (Horizontal, Vertical, Outside Vent Member)
- [ ] <u>U</u>tility control
- [ ] <u>S</u>alvage
- [ ] <u>O</u>verhaul
- [ ] Utilize <u>TIC</u>'s
- [ ] <u>Co</u>ordinate with engine companies (and other companies on scene)
- [ ] Provide periodic <u>P</u>rogress reports

7. **IMS** – U SIL FLOP SR

- ☐ Unified Command Staff (should be considered)
- ☐ Safety officer (to ensure a safe working environment)
- ☐ Information officer (to communicate with the media and concerned citizens)
- ☐ Liaison officer (to communicate with other agencies & ensure smooth operations)
- ☐ Finance officer/section
- ☐ Logistics officer/section
- ☐ Operations officer
- ☐ Planning officer/section
- ☐ Staging officer
- ☐ Rehabilitation officer
- ☐ Establish Divisions/Groups
- ☐ Assign division supervisors

8. **Benchmarks** – Time, Progress, RER/LIP

- ☐ Time Management (Progress reports every 10-15 min)
- ☐ RER/LIP

9. **Under Control** – A SOS PAR ICS DO

- ☐ <u>A</u>nnounce via radio
- ☐ <u>S</u>econdary Searches
- ☐ <u>S</u>alvage and <u>O</u>verhaul responsibilities
- ☐ <u>PAR</u>
- ☐ <u>I</u>nvestigation Unit
- ☐ <u>C</u>O levels

- Secure the Building
- Demobilize
- Turn the building over to the Owner

10. **Terminating/Transferring Command** – TD DOP

- Transfer or Terminate
- Debriefing
- Document/reports
- Offer CISD
- PIA

The Incident Command System (ICS)

BONUS!

There are many benefits to the ICS, but some have argued against it throughout the years. Below are some of those false arguments, followed by a list of benefits of the ICS.

- <u>False arguments against ICS</u>

  - ➤ The ICS is too complex and house fires are not complex.

  - ➤ House fires are short term incidents.

  - ➤ Assigning division/group supervisors takes their hands off the work.

  - ➤ The ICS is check-box firefighting.

  - ➤ It takes the critical thinking out of command.

  - ➤ One size does not fit all.

  - ➤ We tried using ICS before and it didn't work.

- <u>Benefits of the ICS</u>

  - ➤ Improves risk assessment throughout the incident.

  - ➤ Keeps your span of control minimized.

  - ➤ Improves accountability.

  - ➤ Improves communications.

  - ➤ Improves safety.

  - ➤ Matches/forecasts the intensity of the incident (keeps you ahead of the game).

Photo By: Ron Jeffers

You are a newly appointed officer assigned as the tour commander for the shift. At 5:30 PM on a cold night in March, an alarm is transmitted for a reported fire on the first floor of a three-story, ordinary dwelling on Alpine Drive. One alarm has been dispatched to the scene.

The structure is approximately 20' X 60' attached on both sides. On the C side of the building are similar structures separated by narrow fenced-in yards. Before you arrive on scene, you can see a dark smoke plume. The dispatcher informs you they have received several calls by people passing by, on their way home from work.

Upon arrival, Engine 1 reports smoke coming from one window on the first floor. A male occupant from the third floor exits the building saying there is moderate to thick smoke in the second and third floor hallway and he heard the faint cry of woman calling for help as he made his way down the stairs.

As you arrive on scene, Battalion 1 has established command and directed Engine 1 to take a 1 ¾-inch line to the first floor via the front door. You exit your car and walk up to the Battalion Chief. The wind is blowing 12 MPH from the east, towards the C side of the structure.

Scenario Question

What actions would you take as the incident commander, from the initial alarm until termination of the incident?

*Include details on what assignments you would give to the on scene units.

13

1. **En route** – RPM

   ☐ Monitor Radio Reports

   ☐ Review Pre-incident plans

   ☐ Multi-sided view

2. **Upon Arrival** – Position IC CAR PPE

   ☐ Position vehicle properly

   ☐ Face-to-face with current IC

   ☐ Conditions, Actions, Resources (CAR)

   ☐ Assume Command

   ☐ Re-assign IC to operations or interior division

   ☐ Ensure all members are wearing appropriate PPE and respiratory protection

3. **Initial Radio Report** – FM Radio

   ☐ Establish Tactical Radio Frequency.

   ☐ Determine/Announce Operational Mode: (Offensive).

   ☐ IRR: Floors and construction type, Fire conditions, size up concerns, name and Location of the CP (Alpine Drive Command, A/B side of the building).

4. **Size-Up** – COAL TWAS WEALTHS

   ☐ Construction: ordinary, attached dwellings, potential for fire spread – must open bulkhead door to vent.

   ☐ Apparatus/staffing: not sufficient. Call for a second (possible third) alarm.

   ☐ Life hazard: high, due to the type of occupancy.

   ☐ Street conditions: may be busy due to the time of day – traffic.

- Weather: cold, 12 MPH wind toward the C side – will affect fire spread.

- Exposures: must protect exposures on both sides and check for extension.

- Area: check accessibility and conditions in the rear. (You may need to raise the aerial ladder in order to do so… or, if possible, raise ladder to the one-story building on the B side to obtain visibility in the rear.)

- Location/Extent of fire: fire will spread via interior stairs, possible light and air shafts, and pipe chases.

- Time of day: kids/parents may be home from school and work.

5. **Resources** – A RULES WAR+

- <u>A</u>dditional alarms – 2nd on scene, 3rd in staging – to ensure adequate personnel

- <u>R</u>apid Intervention Crew (RIC) - for firefighter safety

- <u>U</u>tility Companies (gas, electric, water) - for utility control

- <u>L</u>aw enforcement - for traffic/crowd control

- <u>E</u>MS (BLS/ambulance, ALS/paramedic) - for patient triage and treatment

- <u>S</u>afety Officer (SO) - for scene safety

- <u>W</u>ater Supply Officer (WSO) - to ensure adequate water

- <u>A</u>ccountability Officer - to track and account for all on scene personnel

- <u>R</u>ehab Unit – for rest and rehabilitation

- Red Cross - for displaced occupants,

- Dept. Public Works (for sand and salt) if freezing temperatures

- Rescue Company – due to high life hazard (calls for help)

6. **Strategy / Tactics** - List the objectives and assign companies

Primary objectives include:

- Locate/Contain the fire

- Remove civilians from danger

- Ventilation the building

Engine Company Ops:

- Position the apparatus past the building for fire attack, leave room in front for the ladder company.

- Establish a primary (and secondary) water supply.

- Initiate attack .

- Choose appropriate size hose lines (1 ¾" min.).

- Advance and position hose lines.

  - Line 1 – interior stairs of the fire building

  - Line 2 – back up first, or floor above

  - Line 3 – floor above fire, or Interior stairs of Exp. D

  - Line 4 – Interior stairs of Exp. D, or exposure B

- Locate, Confine, & Extinguish the fire.

- Protect life hazards by positioning lines in between fire and threatened life.

- Protect the exposures with charged hose lines.

- Supply auxiliary appliances (if there is a sprinkler or standpipe system) .

- Utilize TIC's to assist with searches.

- Coordinate with ladder companies (and other companies on scene).

- Provide periodic progress reports to the IC.

Ladder Company Ops: LOVERS-UPS (TIC-COP)

- Position apparatus in front of building for aerial use – extend to the roof.

- Raise and position Ladders – raise ground ladder to second floor for secondary egress (secure ladders on windy days).

- Force Entry wherever needed in order to access all areas of structure.

- Primary and Secondary Searches of all floors (buildings) must be performed.

- Rescue Operations are priority #1.

- Ventilation (Horizontal, Vertical, and Outside Vent).

- Utility control (Shut gas and electric to fire building).

- ☐ <u>S</u>alvage techniques should be used whenever possible.

- ☐ <u>O</u>verhaul thoroughly to ensure there is no fire extension.

- ☐ Utilize <u>TIC's</u> throughout the operation.

- ☐ <u>C</u>oordinate with engine companies (and other companies on scene).

- ☐ Provide periodic <u>P</u>rogress reports.

7. **<u>IMS</u>** – U SIL FLOP SR

- ☐ Unified Command Staff (not necessary)

- ☐ Safety officer (to ensure a safe working environment)

- ☐ Information officer (to communicate with the media and concerned citizens)

- ☐ Liaison officer (to communicate w/ other agencies & ensure smooth operations)

- ☐ Operations officer should be assigned (consider first or second arriving BC)

- ☐ Also assign:

  - ▪ Staging officer

  - ▪ Rehabilitation officer

  - ▪ Interior, Roof and Rescue Divisions and/or Groups

  - ▪ Division supervisors

8. **<u>Benchmarks</u>** – Time, Progress, RER/LIP

- ☐ Time Management (Progress reports every 10-15 min)

- ☐ RER/LIP

9. **<u>Under Control</u>** – A SOS PAR ICS DO

- ☐ <u>A</u>nnounce via radio

- ☐ <u>S</u>econdary Searches

- ☐ <u>S</u>alvage and <u>O</u>verhaul

- ☐ <u>PAR</u>

- ☐ Investigation Unit
- ☐ CO levels
- ☐ Secure the Building
- ☐ Demobilize
- ☐ Turn the building over to the Owner

10. Terminating/Transferring Command – TD DOP

- ☐ Transfer or Terminate
- ☐ Debriefing
- ☐ Document/reports
- ☐ Offer CISD
- ☐ PIA

Clarifying Statements

In some of the sections in the answer key I provide clarifying statements. For example, under "resources" look at the underlined words:

- ☐ <u>A</u>dditional alarms – 2nd on scene, 3rd in staging – <u>to ensure adequate personnel</u>.

- ☐ <u>R</u>apid Intervention Crew (RIC) - <u>for firefighter safety.</u>

- ☐ <u>U</u>tility Companies (gas, electric, water) - <u>for utility control.</u>

In other sections, such as "under control", I did not use clarifying statements, only the bulleted points that need to be addressed:

- ☐ <u>A</u>nnounce via radio

- ☐ <u>S</u>econdary Searches

- ☐ <u>S</u>alvage and <u>O</u>verhaul

Although I didn't write them down, clarifying statements should always be used. For example:

- ☐ <u>A</u>nnounce via radio so all personnel are aware of the change in status.

- ☐ Ensure <u>S</u>econdary Searches are completed.

- ☐ Conducted <u>S</u>alvage and <u>O</u>verhaul operations to prevent any unnecessary damage or extension.

Clarifying statments will help you come across in a well-organized, professional manner.

It's 7:55 AM on a Tuesday morning. You are the Tour Commander on duty. Your department is dispatched to a report of smoke coming from the area of Main Street on the south end of town. The smoke condition was reported by law enforcement officials out on patrol. You arrive at 123 Main Street moments after the Battalion Chief, who reported heavy smoke coming from the attached garage of a 2,800 square foot home. There is an interior door connecting the garage to the kitchen of the house.

A steady wind is blowing 10 MPH to the west. Across town, one engine and a ladder company are picking up and repacking their hose after an earlier fire. Another engine company is conducting a carbon monoxide investigation at a day care center. Your response on the first alarm is Engine 1, Engine 2, Ladder 1, a BC and yourself.

Scenario Question

As the Incident Commander, what actions would you take from the initial call until the termination of command?

*When you give your answer, describe how you would coordinate fire attack and suppression with ladder company duties.

1. **En route** – RPM

   ☐ Monitor <u>R</u>adio Reports

   ☐ Review <u>P</u>re-incident plans

   ☐ <u>M</u>ulti-sided view

2. **Upon Arrival** – Position IC CAR PPE

   ☐ Position vehicle properly

   ☐ Face-to-face with current IC

   ☐ Incident Debriefing - Conditions, Actions, Resources (CAR)

   ☐ Assume Command

   ☐ Re-assign IC to interior division commander

   ☐ Ensure all members are wearing appropriate PPE and respiratory protection

3. **Initial Radio Report** – FM Radio

   ☐ Establish Tactical Radio <u>F</u>requency.

   ☐ Determine/Announce Operational <u>M</u>ode: (Offensive).

   ☐ IRR: Floors and construction type, Fire conditions, size up concerns, name and location of the CP (Main Street Command, A/B side of the building)

4. **Size-Up** – COAL TWAS WEALTHS

   ☐ Construction: wood-frame construction with an attached garage creates life hazard, fuel source.

   ☐ Occupancy: garages usually have vehicles, flammables liquids, and hazardous materials.

   ☐ Apparatus/staffing: you will need to call for a second alarm; however, response is delayed so call early, and consider a third alarm for staging.

- [ ] Life hazard: high, due to time of day and type of structure. Also, a vehicle is still in the garage.

- [ ] Weather: wind is consistent, which can cause fire to spread if/when garage doors are open and fire is not extinguished quickly.

- [ ] Exposures: biggest exposure threat is attached residence – kitchen door and room above.

- [ ] Location/Extent of fire: potentially a heavy fire condition due to delayed discovery.

- [ ] Time of day: residents are likely to still be in the house at 0755.

5. **Resources** – A RULES WAR+

- [ ] <u>A</u>dditional alarms – 2nd on scene, 3rd in staging – to ensure adequate personnel – mutual aid may be your second alarm due to busy companies from previous calls.

- [ ] <u>R</u>apid Intervention Crew (RIC) - for firefighter safety

- [ ] <u>U</u>tility Companies (gas, electric, water) - for utility control

- [ ] <u>L</u>aw enforcement - for traffic/crowd control

- [ ] <u>E</u>MS (BLS/ambulance, ALS/paramedic) - for patient triage and treatment

- [ ] <u>S</u>afety Officer (SO) - for scene safety

- [ ] <u>W</u>ater Supply Officer (WSO) - to ensure adequate water

- [ ] <u>A</u>ccountability Officer - to track and account for all on scene personnel

- [ ] <u>R</u>ehab Unit – for rest and rehabilitation

- [ ] Red Cross - for displaced occupants,

- [ ] Rescue Company – for possible trapped occupants

6. **Strategy / Tactics** - List the objectives and assign companies

Primary objectives include:

- [ ] Locate/Contain the fire

- [ ] Protect the residents in home

- ☐ Remove civilians from danger

- ☐ Ventilate the building

- ☐ Check for fire extension

Engine Company Ops:

- ☐ Position apparatus past building for fire attack, leave room in front for the ladder company.

- ☐ Establish a primary (and secondary) water supply.

- ☐ Initiate attack.

- ☐ Choose appropriate size hose lines (1 ¾" line for speed and mobility).

- ☐ Advance and position hose lines.

  - ▪ Line 1 – through house, to interior door. Keep door closed, protect life/ prevent spread.

  - ▪ Line 2 – through the garage doors to extinguish the fire.

  - ▪ Line 3 – back up the fire attack line.

  - ▪ Line 4 – possibly inside the structure to back up interior line, or floor above.

  - ▪ Note: If you determine there is no interior life hazard, you may choose to enter through the interior door and use the garage doors to ventilate.

- ☐ Do not enter the garage without a charged hose line.

- ☐ When passing under a roll-down garage door, secure it in an open position by wedging the tip of the pike pole into the track or clamping a vice grip on the track to prevent the door from lowering.

- ☐ If possible, chock one wheel of the vehicle on fire to prevent movement.

- ☐ Locate, Confine, & Extinguish the fire.

- ☐ If the hose line will not extinguish the fire, consider using foam or a dry chemical extinguisher.

- ☐ Protect life hazards by keeping the interior door shut.

- ☐ Protect any threatened exposures with charged hose lines.

- ☐ If burning flammable liquid is flowing out of the garage, cover it with foam, dike it, and check threatened drains and sewers for flammable vapors.

- ☐ Utilize TIC's to assist with searches.
- ☐ Check for extension through open channels that can provide a path of travel .
- ☐ If you suspect fire extension in multiple areas, stretch additional lines.
- ☐ Coordinate with ladder companies (and other companies on scene).
- ☐ Provide periodic progress reports to the IC.

Ladder Company Ops: LOVERS-UPS (TIC-COP)

- ☐ <u>P</u>osition apparatus in front of building.
- ☐ Raise and position <u>L</u>adders – raise ground ladder to second floor for secondary egress (secure ladders on windy days).
- ☐ Force <u>E</u>ntry wherever needed to access all areas of structure.
- ☐ Primary and Secondary <u>S</u>earches of all floors (buildings) must be performed:
  - search the vehicle and trunk (afterwards, secure the trunk in a closed position).
  - search above the garage.
  - search the garage for flammables and hazardous materials.
- ☐ <u>R</u>escue Operations are priority #1.
- ☐ <u>V</u>entilation (Horizontal, using an Outside Vent Member to remove the garage windows).
- ☐ <u>U</u>tility control (Shut gas and electric to the fire building).
- ☐ <u>S</u>alvage techniques should be used whenever possible.
- ☐ <u>O</u>verhaul thoroughly to ensure there is no fire extension through open channels that can provide a path of travel such as damaged doors and walls, ductwork, poor workmanship, pipe chases and recesses.
- ☐ Utilize <u>TIC</u>'s throughout the operation.
- ☐ <u>C</u>oordinate with engine companies (and other companies on scene).
- ☐ Provide periodic <u>P</u>rogress reports.
- ☐ Clean up the scene (use absorbent or sand for small spills, sweep glass, and re-

move any large or sharp debris from the ground).

- ☐ Expose the least amount of firefighters to the garage area.

7. **IMS** – U SIL FLOP SR

- ☐ Unified Command Staff (not necessary)
- ☐ Safety officer (to ensure a safe working environment)
- ☐ Information officer (to communicate with the media and concerned citizens)
- ☐ Liaison officer (to communicate w/ other agencies & ensure smooth operations)
- ☐ Operations officer should be assigned (consider first or second arriving BC)
- ☐ Also assign:
  - Staging officer
  - Rehabilitation officer
  - Interior, Roof and Rescue Divisions and/or Groups
  - Division supervisors

8. **Benchmarks** – Time, Progress, RER/LIP

- ☐ Time Management (Progress reports every 10-15 min)
- ☐ RER/LIP

9. **Under Control** – A SOS PAR ICS DO

- ☐ <u>A</u>nnounce via radio
- ☐ <u>S</u>econdary Searches
- ☐ <u>S</u>alvage and <u>O</u>verhaul
- ☐ <u>PAR</u>
- ☐ <u>I</u>nvestigation Unit - If this is a suspected crime scene, reserve the evidence.
- ☐ <u>C</u>O levels
- ☐ <u>S</u>ecure the Building

- ☐ <u>D</u>emobilize

- ☐ Turn the building over to the <u>O</u>wner

10. **<u>Terminating/Transferring Command</u>** – TD DOP

- ☐ <u>T</u>ransfer or Terminate

- ☐ <u>D</u>ebriefing

- ☐ <u>D</u>ocument/reports

- ☐ <u>O</u>ffer CISD

- ☐ <u>P</u>IA

Garage Fire Tip

BONUS!

A vehicle burning inside a garage is a structure fire and should be treated as such. Garage fires are complex and pose all the threats of a structure fire, car fire, and a haz-mat incident due to stored goods such as pool and lawn chemicals, paints, and other flammables.

Fire Scenario #3 - Church Fire

Photo By: Ron Jeffers

It's after midnight. You are returning from a fire that occurred earlier in the evening. Two companies are still at the scene wrapping up, when an alarm is transmitted for a fire at St. Thomas' Church. Two engines, one ladder company and a battalion chief are also dispatched along with you–the tour commander on duty.

St. Thomas' Church is an old, gothic style structure in your community. The building is approximately 120' x 200', with a large open basement, a slate roof, and an attached rectory. There are doors on three sides of the basement, which is partially below grade. In addition to the rectory on the C side, there are two-story apartments on both the B and D side of the church.

The first arriving Engine and the Battalion Chief arrive on scene a few minutes before you and report smoke coming from the rear of the church. It appears to be coming from the first floor or the basement. Ladder one and the second engine arrive on scene with you. It's 65° and the wind is blowing approximately 15 mph toward the Northeast.

Scenario Question

As the Incident Commander, what actions would you take from the initial call until the termination of command?

Fire Scenario #3 - Answer Key

1. **En route** – RPM

   ☐ Monitor <u>R</u>adio Reports

   ☐ Review <u>P</u>re-incident plans

   ☐ <u>M</u>ulti-sided view

2. **Upon Arrival** – Position IC CAR PPE

   ☐ Position vehicle properly

   ☐ Face-to-face with current IC

   ☐ Incident Debriefing - Conditions, Actions, Resources (CAR)

   ☐ Assume Command

   ☐ Re-assign IC to operations or interior division

   ☐ Ensure all members are wearing appropriate PPE and respiratory protection

3. **Initial Radio Report** – FM Radio

   ☐ Establish Tactical Radio <u>F</u>requency.

   ☐ Determine/Announce Operational <u>M</u>ode: (Offensive).

   ☐ IRR: Floors and construction type, Fire conditions, size up concerns, name and Location of the CP (St Thomas Command, C/D side of the building)

4. **Size-Up** – COAL TWAS WEALTHS

   ☐ Construction:

   - The slate tile roofs will be difficult (maybe impossible) to access and ventilate.

   - The attached rectory will need to be searched and evacuated.

   - Large fire potential, use 2-1/2" hose lines.

30

- Large open spaces: utilize TIC's and search ropes.

- Concealed spaces throughout = easy fire travel.

- FF's may find it challenging to access the lofts, ceiling spaces, & basement areas.

☐ Occupancy: There will be numerous priceless contents within the Church. Special emphasis should be placed on efficient salvage and overhaul operations.

☐ Apparatus/staffing: Not sufficient, need additional alarms.

☐ Life hazard:

- Priests in rectory. Search and evacuate. Account for all members.

- Firefighters being physically taxed by operations (ventilation, search, stretching lines, eventual collapse potential).

☐ Water supply: Delayed discovery will require more water.

☐ Exposures: The attached rectory and apartments on B and D sides must be evacuated and protected. Assign personnel to control utilities.

☐ Location/Extent of fire: This is a possibly basement fire, expect fire travel. You will need additional personnel for long hose line stretches.

☐ Time of day: Delayed alarms, and advanced fire conditions. This fire will require more water and staffing.

5. **Resources** – A RULES WAR+

☐ <u>A</u>dditional alarms – Min. 2nd and 3rd to the scene – to ensure adequate personnel

☐ <u>R</u>apid Intervention Crew (RIC) - for firefighter safety

☐ <u>U</u>tility Companies (gas, electric, water) - for utility control

☐ <u>L</u>aw enforcement - for traffic/crowd control

☐ <u>E</u>MS (BLS/ambulance, ALS/paramedic) - for patient triage and treatment

☐ <u>S</u>afety Officer (SO) - for scene safety

☐ <u>W</u>ater Supply Officer (WSO) - to ensure adequate water

☐ <u>A</u>ccountability Officer - to track and account for all on scene personnel

☐ <u>R</u>ehab Unit – for rest and rehabilitation (call early due to previous fire)

- ☐ Rescue Company – due to high life hazard (priests in rectory)

6. **<u>Strategy / Tactics</u>** - List the objectives and assign companies

Primary objectives include:

- ☐ Locate/Contain/Extinguish the fire

- ☐ Remove priests/civilians from danger

- ☐ Ventilate the building

Engine Company Ops:

- ☐ Position apparatus past the building for fire attack, leave room in front for the ladder company.

- ☐ Establish a primary (and secondary) water supply.

- ☐ Initiate attack.

- ☐ Choose appropriate size hose lines (2 ½").

- ☐ Advance and position hose lines:
  - ▪ Line 1 – to the fire
  - ▪ Line 2 – back up first line
  - ▪ Line 3 – floor above the fire
  - ▪ Line 4 – protect downwind exposure

- ☐ Locate, Confine, & Extinguish the fire

- ☐ Protect life hazards by positioning lines in between fire and life.

- ☐ Protect exposures with charges hose lines.

- ☐ Supply auxiliary appliances (if there is a sprinkler or standpipe system).

- ☐ Utilize TIC's to assist with searches.

- ☐ Coordinate with ladder companies (and other companies on scene).

- ☐ Provide periodic progress reports to the IC.

Ladder Company Ops: LOVERS-UPS (TIC-COP)

- Position apparatus in front of the building for aerial use – to ventilate stained glass window, if needed – Stay out of collapse zone.

- Raise and position Ladders – raise ground ladder to the second floor or rectory for secondary egress.

- Force Entry wherever needed to access all areas of structure.

- Primary and Secondary Searches of all floors (church, rectory and threatened exposure).

- Rescue Operations are priority #1 (Evacuate church, rectory and threatened exposure).

- Ventilation (Horizontal, Vertical, and Outside Vent Member).

- Utility control (Shut gas and electric to fire building).

- Salvage techniques should be used to preserve valuable/irreplaceable contents.

- Overhaul thoroughly to ensure there is no fire extension.

- Utilize TIC's throughout the operation.

- Coordinate with engine companies (and other companies on scene).

- Provide periodic Progress reports.

Additional:

- All personnel, Preserve evidence.

- Prepare for possible defensive operations (See answer key for Fire Scenario #10).

7. **IMS** – U SIL FLOP SR

- Unified Command Staff (should be considered)

- Safety officer (to ensure a safe working environment) May assign two SO's for large incidents

- Information officer (to communicate with the media and concerned citizens)

- Liaison officer (to communicate w/ other agencies & ensure smooth operations)

- Finance officer/section

- ☐ Logistics officer/section

- ☐ Operations officer – assign a Battalion Chief

- ☐ Planning officer/section

- ☐ Staging officer (designate an area nearby for staging)

- ☐ Rehabilitation officer (designate an area nearby for staging)

- ☐ Establish Divisions/Groups (Basement division, vent group, evacuation group, etc)

- ☐ Assign division supervisors

8. **Benchmarks** – Time, Progress, RER/LIP

- ☐ Time Management (Progress reports every 10-15 min)

- ☐ RER/LIP

9. **Under Control** – A SOS PAR ICS DO

- ☐ Announce via radio

- ☐ Secondary Searches

- ☐ Salvage and Overhaul

- ☐ PAR

- ☐ Investigation Unit

- ☐ CO levels

- ☐ Secure the Building

- ☐ Demobilize

- ☐ Turn the building over to the Owner

10. **Terminating/Transferring Command** – TD DOP

- ☐ Transfer or Terminate

- ☐ Debriefing

- ☐ <u>D</u>ocument/reports
- ☐ <u>O</u>ffer CISD
- ☐ <u>P</u>IA

Church Fires Present Life Hazards 24/7 BONUS!

Experienced chief officers and IC's will often tell you that fires in churches come in two sizes—small and very large, with a quick transition between the two. If that wasn't enough, you can expect the churches of today to take on many additional roles other than Sunday service. From day care centers, homeless shelters, and banquet halls, to parochial schools and food kitchens, the additional occupancy concerns can take place any day of the week, at any time of the day or evening. A fire involving a church can result in a serious loss of life and property.

You are a newly appointed officer assigned as the tour commander because your superior is on temporary sick leave. It's a Sunday afternoon in September when a man calls 911 and reports that flames are coming from the basement window of his neighbor's home at 33 Oak Street. He thinks a family of four – husband, wife and two children under the age of 5 – may still be inside.

You arrive on scene simultaneously with Engine 1 and Ladder 1 and confirm this is a basement fire. Flames are visible from the window on the A/B side of this one-family, two-story wood frame dwelling. Another neighbor tells you he thinks the family went on vacation for the weekend. Over the radio, you hear Engine 2 report that their apparatus was involved in a vehicle accident and they are going to be delayed. You also hear an unfamiliar popping sound coming from the basement.

Photo By: Ron Jeffers

You look around and realize there are no serious exterior exposure threats, as long as the fire is contained to the basement. There are two cars and one childs bike in the driveway. A crowd is beginning to form. One man is coming from the rear yard of the building. He is holding a phone in his hand, which he was using to film the fire, and now the firefighters, as they arrive on scene.

There is no wind or weather concern on this day.

Scenario Question

As the Incident Commander, what actions would you take from the initial call until the termination of command?

Within your answer, list some common concerns and hazards that are usually found in basements and cellars.

1. **En route** – RPM

    ☐ Monitor <u>R</u>adio Reports

    ☐ Review <u>P</u>re-incident plans

    ☐ <u>M</u>ulti-sided view

2. **Upon Arrival** – Position IC CAR PPE

    ☐ Position vehicle properly

    ☐ Establish Command

    ☐ Determine Conditions, Actions, Resources (CAR)

    ☐ Ensure all members are wearing appropriate PPE and respiratory protection

    ☐ Quickly complete a 360° evaluation of the structure prior to entering. Look for fire, windows, exterior doors, bilco doors and any hazards.

    ☐ Try to determine if the man with the camera/phone lives there. Direct him away from the property, but not away from the area (point him out to the police when they arrive).

3. **Initial Radio Report** – FM Radio

    ☐ Establish Tactical Radio <u>F</u>requency.

    ☐ Determine/Announce Operational <u>M</u>ode: (Offensive).

    ☐ IRR: Floors and construction type, Fire conditions, size up concerns, name and Location of the CP (Oak Street Command, A/B side of the building)

4. <u>Size-Up</u> – COAL TWAS WEALTHS

    ☐ Construction: Open interior stairwell, pipe chases, possible balloon frame construction – all would cause rapid fire spread.

    ☐ Occupancy: The following hazards/concerns are usually present in basements of residential homes:

- Flammable liquids: gasoline, paint solvents, fuel oil. (May be the cause of the popping sound)

- Heating units, fueled by natural gas and oil. Attempt to shut down the heating unit by using the remote switch on top of the basement stairs.

- Intense heat due to lack of ventilation in the closed up area.

- Limited entry because most basements have only one entrance.

☐ Apparatus/staffing: Not sufficient, need an additional Engine company, call for a second alarm.

☐ Life hazard: Must assume all residential occupancies present a life hazard. Cars in the driveway may be a clue.

☐ Location/Extent of fire: Basement, expect upward fire travel. Ypu will need additional personnel for long hose line stretches.

☐ Time of day: Sunday afternoon in residential structures present probable life hazards.

☐ Special Considerations: The man with the video/phone needs to be questioned by fire prevention and law enforcement personnel.

5. **Resources** – A RULES WAR+

☐ <u>A</u>dditional alarms – 2nd to the scene, 3rd in staging – to ensure adequate personnel

☐ <u>R</u>apid Intervention Crew (RIC) - for firefighter safety

☐ <u>U</u>tility Companies (gas, electric, water) - for utility control

☐ <u>L</u>aw enforcement - for traffic/crowd control – ALSO: question man with camera, and dispatch a unit to scene of accident to check on status of Engine 2

☐ <u>E</u>MS (BLS/ambulance, ALS/paramedic) - for patient triage and treatment

☐ <u>S</u>afety Officer (SO) - for scene safety. You may also assign an additional SO to the rear (Side C)

☐ <u>W</u>ater Supply Officer (WSO) - to ensure adequate water

☐ <u>A</u>ccountability Officer - to track and account for all on scene personnel

☐ <u>R</u>ehab Unit – for rest and rehabilitation

☐ Red Cross – possibly – for displaced occupants

6. **Strategy / Tactics** - List the objectives and assign companies

Primary objectives include:

☐ Locate/Contain the fire; Remove civilians from danger; Ventilation the building; etc.

Engine Company Ops:

☐ Position apparatus past building for fire attack, leave room in front for the ladder company.

☐ Establish a primary (and secondary) water supply .

☐ Initiate attack.

☐ Choose appropriate size hose lines (1 ¾" min.).

☐ Advance and position hose lines.

➢ If there is <u>NO</u> exterior basement door:

▪ Advance the first line (1-3/4") to interior stairwell, close door, protect.

▪ If fire is in the incipient stage with no life hazard; advance, and extinguish.

▪ If advanced, close the door, and evacuate occupants to safety.

▪ Advance a line into the basement <u>only</u> when the backup line is in place.

▪ If the position becomes untenable, officer must notify the IC so companies above the fire can evacuate before the engine company withdraws.

➢ If there is an <u>Exterior Basement Door</u>:

▪ 1st Line, interior stairs, close door, protect life.

▪ 2nd Line, back up first line.

▪ 3rd Line, enter through exterior door to LCE.

▪ *Balloon frame – need to stretch a line to the top floor/attic*

☐ Locate, Confine, & Extinguish the fire.

☐ Protect life hazards by positioning lines in between fire and life.

- Protect the exposures with charged hose lines.

- Supply auxiliary appliances (if there is a sprinkler or standpipe system).

- Utilize TIC's to assist with searches.

- Coordinate with ladder companies (and other companies on scene).

- Provide periodic progress reports to the IC.

- Additional Eng. Co. Tips:

  - Stretch a line long enough to reach the far corner.

  - Charge the line and bleed the air out before entering the basement.

  - When advancing hose lines down the interior stairs, move quickly and stay to the bearing wall if possible.

  - Utilize straight stream, fog will produce steam.

  - Remember: When an exterior company hits the fire, the company on top of the stairs is in a compromising position.

  - Never enter the basement without the protection of a hose line.

Ladder Company Ops: LOVERS-UPS (TIC-COP)

- Position apparatus in front of building for possible aerial use and easy access of ground ladders.

- Raise and position Ladders – raise ground ladders to second floor for secondary egress (residents may be trapped by smoke).

- Force Entry wherever needed to access all areas of the structure.

- Primary and Secondary Searches of all floors must be performed.

  - Evening hours, start with bedrooms – Vent in the process (VES).

  - Simultaneously search for fire extension (walls, attic, cockloft) use TIC.

  - Check baseboards and walls directly over the fire for extension.

- Rescue Operations are priority #1 – evacuate and account for all residents.

- Ventilation (Horizontal, Use an Outside Vent Member) will need personnel to access the attic, and possibly open the roof in balloon frame construction.

- Utility control (Shut gas and electric to fire building) If you cannot do so from the basement, shut utilities from the street.

- Salvage techniques should be used whenever possible.
- Overhaul thoroughly to ensure no fire extension.
- Utilize TIC's throughout the operation.
- Coordinate with engine companies (and other companies on scene).
- Provide periodic Progress reports.

7. **IMS** – U SIL FLOP SR

- Safety officer (to ensure a safe working environment)
- Information officer (to communicate with the media and concerned citizens)
- Liaison officer (to communicate w/ other agencies & ensure smooth operations)
- Operations officer should be assigned (consider first or second arriving BC)
- Also assign:
    - Staging officer, Designate staging area
    - Rehabilitation officer, designate rehab area
    - Basement, Ventilation and Evacuation Divisions and/or Groups
    - Division supervisors

8. **Benchmarks** – Time, Progress, RER/LIP

- Time Management (Progress reports every 10-15 min) - RER/LIP

9. **Under Control** – A SOS PAR ICS DO

- Announce via radio
- Secondary Searches
- Salvage and Overhaul
- PAR
- Investigation Unit (be sure to interview man with camera)

- □ <u>C</u>O levels

- □ <u>S</u>ecure the Building

- □ <u>D</u>emobilize

- □ Turn the building over to the <u>O</u>wner

10. **<u>Terminating/Transferring Command</u>** – TD DOP

- □ <u>T</u>ransfer or Terminate

- □ <u>D</u>ebriefing

- □ <u>D</u>ocument/reports

- □ <u>O</u>ffer CISD

- □ <u>P</u>IA

Unaccounted for Residents

BONUS!

In this scenario, you were given conflicting reports as to the whereabouts of the family. If residents are unaccounted for, we are taught in the fire service to always assume someone is still inside.

It's 5:30 in the morning on December 28th. Your community, which has sent a reverse 911 weather advisory requesting that people stay off the roads and in their homes if possible, has just been hit with more than twelve inches of snow in the last 48 hours.

An employee at one of the two nursing homes in your community called the fire department stating there is a smoke condition and strange burning odor coming from a medical supply closet in the center of the second floor, behind the nurse's desk. The building, located at 200 Main Street, is a large four story, non-combustible structure. Unlike the other nursing home, which is partially occupied, this one is a fully occupied, 100-bed nursing home taking up three quarters of the building, with 50 elderly residents on both the second and third floor. The first floor is a small lobby and staff quarters, the fourth floor is under construction. The other 1/4 portion of the building is a newly constructed drug and alcohol rehabilitation center, with separate utilities and HVAC system. Each portion of the building has an elevator and two stair-wells. The elevator in the nursing home portion of the building was temporarily placed out of service yesterday.

Luckily, the nursing home and your fire station are on the same road, which has been plowed and salted about sixty-minutes earlier. When you arrive on scene, a nurse meets you in the lobby and confirms an odd odor and moderate smoke condition coming from the second floor, which is the floor where the occupants with more severe respiratory and medical conditions stay. Some of the patients are on dialysis. The nurse is panicking, stating she has no place to relocate the patients on that floor. The fire seems confined to the second floor closet. The smoke and odor seems to only affect the second floor. This building is supplied with a sprinkler and standpipe system. You discover the sprinkler in the closet has activated. Water is dripping from the first floor ceiling fixtures. It's clear something is burning from the odor, but you cannot identify the smell. Early reports inform you the odor is in every room on the second floor, even though the doors are shut. Shortly after your arrival, a news television crew that was filming storm related footage pulls on scene and the reporter begins asking questions to firefighters.

Scenario Question

As the Incident Commander, what actions would you take from the initial call until the termination of command?

Fire Scenario #5 - Answer Key

1. **En route** – RPM

   ☐ Monitor <u>R</u>adio Reports

   ☐ Review <u>P</u>re-incident plans

   ☐ <u>M</u>ulti-sided view

2. **Upon Arrival** – Position IC CAR PPE

   ☐ Position vehicle properly

   ☐ Establish Unified Command (along with law enforcement, OEM, EMS, and the head nurse)

   ☐ Determine Conditions, Actions, Resources (CAR)

   ☐ Ensure all members are wearing appropriate PPE and respiratory protection

   ☐ Establish communications with the head nurse or a representative, and request a list of medications that are in the supply closet (MSDS)

3. **Initial Radio Report** – FM Radio

   ☐ Establish Tactical Radio <u>F</u>requency.

   ☐ Determine/Announce Operational <u>M</u>ode: (Offensive).

   ☐ IRR: Floors and construction type, Fire conditions, size up concerns, name and Location of the CP (Main Street Command, Located in the lobby of the building)

4. **Size-Up** – COAL TWAS WEALTHS

   ☐ Apparatus/staffing: not sufficient. You will need additional staffing to handle suppression, ventilation, and patient relocation.

   ☐ Life hazard: very high life hazard. Elderly patients will need assistance, many of them have respiratory issues. No room to relocate them to another floor.

- ☐ Auxiliary appliances: Sprinkler System and Standpipe will need to be supplied.

- ☐ Street conditions: icy, need to be plowed and salted.

- ☐ Weather: very cold, will be challenging for patients who are being transported.

- ☐ Location/Extent of fire: Unknown contents, confined to closet, but in the middle of an occupied floor.

- ☐ Time of day: big factor due to possible late discovery, sleeping occupants, and light on staff personnel.

- ☐ Special considerations: The reporter on scene will need to be dealt with.

5. **Resources** – A RULES WAR+

- ☐ <u>A</u>dditional alarms – 3rd Alarm – to ensure adequate personnel to remove patients

- ☐ <u>R</u>apid Intervention Crew (RIC) - for firefighter safety

- ☐ <u>U</u>tility Companies (gas, electric, water) - for utility control

- ☐ <u>L</u>aw enforcement - for traffic/crowd control

- ☐ <u>E</u>MS (BLS/ambulance, ALS/paramedic) - for patient triage and treatment

- ☐ <u>S</u>afety Officer (SO) - for scene safety

- ☐ <u>W</u>ater Supply Officer (WSO) - to ensure adequate water (if needed)

- ☐ <u>A</u>ccountability Officer - to track and account for all on scene personnel

- ☐ <u>R</u>ehab Unit – for rest and rehabilitation

- ☐ Red Cross – possibly – to assist with supplies for displaced residents

- ☐ Haz-Mat specialist, to monitor air quality and investigate the danger concerning the medical supplies within the closet (unusual odor).

6. **Strategy / Tactics** - List the objectives and assign companies

Primary objectives include:

- ☐ Locate/Contain the fire;

- ☐ Remove (relocate) occupants from danger;

- ☐ Ventilate the building; etc.

Strategy includes:

- Direct personnel to fire floor. Have them stretch a line, but keep the closet door shut.

- Work with designated EMS member to ensure they call for heated transportation.

- Work with the nurse (and an OEM coordinator) to see if the other nursing home is available for the 50 patients who need to be relocated.

- Ensure there are enough nurses and sufficient medical supplies at the transfer facility, call for additional companies to go to that location to assist with carrying patients.

- Have EMS & fire personal begin carrying patients down to the lobby to begin relocating.

- Use blankets to shield patients from bad weather.

- Call for trucks to plow and apply salt to a path from one nursing home to the other.

- Assign a victim tracking coordinator. Ensure family members of patients are informed.

- Set up outdoor lighting and salt the sidewalk/road in front of the nursing home to prevent slipping hazards.

- Ensure nurses or ALS personnel travel with patients on dialysis machines.

Engine Company Ops:

- Position apparatus past the building for fire attack, leave room in front for the ladder company.

- Establish a primary (and secondary) water supply.

- Initiate attack.

- Choose appropriate size hose lines.

- Advance and position hose lines:

  - 1st line: 1 ¾" to the medical supply closet

  - 2nd line: back up first with equal size line if needed.

49

- ☐ Locate, Confine, & Extinguish the fire.

- ☐ Protect life hazards by keeping doors shut (closet door should be shut when removing patients).

- ☐ Protect exposures (search floor/room above the closet for fire spread).

- ☐ Supply auxiliary appliances (sprinkler and possibly standpipe system).

- ☐ Utilize TIC's to assist with searches.

- ☐ Coordinate with ladder companies (and other companies on scene).

- ☐ Provide periodic progress reports to the IC.

Ladder Company Ops: LOVERS-UPS (TIC-COP)

- ☐ <u>P</u>osition apparatus in front of building for aerial use – be prepared to extend to the roof.

- ☐ Be prepared to raise and position <u>L</u>adders if needed.

- ☐ Force <u>E</u>ntry wherever needed to access all areas of structure.

- ☐ Primary and Secondary <u>S</u>earches of all floors (buildings) must be performed/ Search the fire floor, the floors above, and rehab center. If other occupants are safe, protect in place.

- ☐ <u>R</u>escue Operations are priority #1.

- ☐ <u>V</u>entilation (Horizontal – mechanical ventilation after evacuation would be best option).

- ☐ <u>U</u>tility control (Shut HVAC system).

- ☐ <u>S</u>alvage techniques should be used whenever possible.

- ☐ <u>O</u>verhaul thoroughly to ensure no fire extension.

- ☐ Utilize <u>TIC</u>'s throughout the operation.

- ☐ <u>C</u>oordinate with engine companies (and other companies on scene).

- ☐ Provide periodic <u>P</u>rogress reports.

7. **IMS** – U SIL FLOP SR

- ☐ Unified Command Staff (Include: Fire, Police, EMS, and a nursing home representative).

- ☐ Safety officer (to ensure a safe working environment).

- ☐ Information officer (to communicate with the media and concerned family members).

- ☐ Liaison officer (to communicate w/ other agencies & ensure smooth operations).

- ☐ Finance officer/section.

- ☐ Logistics officer/section.

- ☐ Operations officer (assign BC).

- ☐ Planning officer/section.

- ☐ Designate staging officer, and area in the lobby. They can also help move patients.

- ☐ Designate Rehabilitation officer and location.

- ☐ Establish ventilation and evacuation Divisions/Groups.

- ☐ Assign division supervisors.

8. **Benchmarks** – Time, Progress, RER/LIP

- ☐ Time Management (Progress reports every 10-15 min)

- ☐ RER/LIP

9. **Under Control** – A SOS PAR ICS DO

- ☐ <u>A</u>nnounce via radio

- ☐ <u>S</u>econdary Searches

- ☐ <u>S</u>alvage and <u>O</u>verhaul

- ☐ <u>P</u>AR

- ☐ <u>I</u>nvestigation Unit

- ☐ <u>C</u>O levels

- ☐ <u>S</u>ecure the Building

- ☐ <u>D</u>emobilize

    ☐ Turn the building over to the <u>O</u>wner

10. **<u>Terminating/Transferring Command</u>** – TD DOP

    ☐ <u>T</u>ransfer or Terminate

    ☐ <u>D</u>ebriefing

    ☐ <u>D</u>ocument/reports

    ☐ <u>O</u>ffer CISD

    ☐ <u>P</u>IA

Sandwich Theory

BONUS!

When practicing fire scenarios, you will begin to realize the benefits of what I call the Sandwich Theory. Simply put, the beginning and end of each scenario is essentially the same (with minor differences) Consider the information provided in the first three and the last three sections of each of the first five scenarios and you will get the idea.

The actions you will take in sections 1-3: En route, Upon Arrival, and Initial Radio report will remain very similar regardless of the fire size and the same can be said regarding the actions you will take in sections 8-10: Benchmarks, Under Control, Terminating/Transferring Command. The main difference between each of these scenarios is the middle. More specifically, the strategy and tactics section - that is where the meat is.

Fire Scenario #6 - Auto Body Shop

Photo By: Ron Jeffers

You are dispatched to a fire in an old auto body shop that has been out of business for more than five years. The structure is surrounded by empty lots and there is no exposure threat. Shortly after arrival, the fire reaches and quickly burns through a portion of the truss roof. You have already chosen a defensive strategy, set up master streams, and are in the process of taking the necessary precautions needed to establish safe work zones. It is 7:45 AM and 31°. The wind is blowing approximately 22 MPH.

Scenario Questions

1. What are your primary safety considerations concerning this type of fire?

2. Name 10 or more signs you would consider when determining collapse potential?

53

Fire Scenario #6 - Answer Key

1. Primary safety considerations include:

   ☐ Truss roof involvement

   ☐ Possible hazardous materials

   ☐ Possible squatters

   ☐ Delayed discovery

   ☐ Heavy fire upon arrival

   ☐ Water supply challenges

   ☐ Smoke travel

   ☐ Strong winds will result in rapid fire spread

   ☐ Personnel accountability

   ☐ Rehabilitation

   ☐ Collapse potential

   ☐ Water runoff, icy parking lot, slipping hazards

2. Signs to consider when determining collapse potential include:

   ☐ Fire size and location

   ☐ Heavy fire for an extended period of time

   ☐ Pieces of the building falling off

   ☐ Cracks in wall

   ☐ Leaning or bowing walls

   ☐ Building age and condition

   ☐ Faulty/poor construction

   ☐ Foundation failure

   ☐ Extraordinary loads

- ☐ Lack of water runoff

- ☐ Sagging floors or beams

- ☐ Spongy roof or floors

- ☐ Previous fires at this location

- ☐ Explosions, flashovers, or backdrafts

- ☐ Water and/or smoke pushing through solid masonry walls

- ☐ Smoke through mortar joints

- ☐ Accidental cutting of structural support members

- ☐ Lightweight construction components

- ☐ Extreme weather conditions

- ☐ Fire reaches the truss roof

- ☐ Unusual noises (creaking)

- ☐ Any combination of causes

Bullet Points

BONUS!

As you read books and study, any time you come across bullet points on any specific topic, highlight them or write them down so you can refer to them often when you study. Doing this will give you a better chance of retaining info you will need to recall at a moment's notice.

Fire Scenario #7 - Scrap Metal Fire

Photo By: Bill Tompkins

It's 6:12 AM on a Wednesday morning. You are dispatched to the report of a large fire at a local scrap yard. While en route, the dispatcher informs all units that he has received several calls regarding this fire. You see the glow of fire from across the town.

When you arrive on scene, due to the lack of light, you have all apparatus drivers set up their portable lighting and spotlights to illuminate the area. You know from past incidents that these types of fires present challenges and concerns unique to a typical structure fire. For example, workers begin showing up on these sites as early as 5:00 AM.

Your local water department had just finished putting in a new hydrant on a large main at the entrance of the site. Your department has not yet tested the hydrant to ensure adequate flow.

Scenario Question

According to the text *Fireground Operational Guides*, what <u>incident specific</u> actions should be taken at scrap metal and junkyard fires?

Fire Scenario #7 - Answer Key

1. Interview employees upon arrival.

    ☐ Determine what is burning and if there is a life hazard.

    ☐ Assign a victim tracking coordinator to account for all employees.

    ☐ Remove employees and civilians from danger.

2. Survey the scene

    ☐ Conduct a thorough size-up; emphasize the following: Apparatus and staffing needs, Apparatus positioning, Life hazards (firefighter safety is a priority), Water supply, Weather conditions, Threatened exposures, and Location and Extent of the fire.

3. Call early for resources

    ☐ This may include additional alarms, law enforcement, emergency medical services (EMS), etc.

    ☐ Notify the Environmental Protection Agency (EPA) or your designated environmental specialist. If they provide you with a case number, include it in your report.

4. Wear full PPE and SCBA

    ☐ Firefighters must take precautions to protect themselves from various hazards including flammable liquids, BLEVE's, burning tires and upholstery, and so on.

    ☐ Monitor air quality and smoke cloud.

5. Select the appropriate size line

    ☐ Use a minimum 2 ½-inch line (If needed, add a length of 1 ¾-inch hose at the end for mobility).

    ☐ Consider using a deck gun or an elevated stream.

> ➤ Large-caliber streams can be used to extinguish and drive a stream into the top of a pile of burning material while keeping firefighters at a safe distance.

☐ Consider covering the area with foam.

☐ Pay attention to water runoff.

6. Overhaul tips include:

   ☐ Use the minimum amount of firefighters necessary to do the job.

   ☐ Do not allow single firefighters to move heavy objects alone.

   ☐ Use streams to move debris and overhaul hydraulically.

   ☐ If available, allow qualified employees to operate heavy equipment (cranes, back loaders, and high-lows) to move objects such as heavy metals and wrecked autos.

7. Establish a fire watch and/or revisit the scene to check for and prevent flare-ups.

   ☐ In daylight hours, perform a complete overhaul with a hose line as heavy materials and wrecked cars are being moved.

Critical Reading BONUS!

The question is simple and direct. You may choose to go through an entire scenario format, but this may be unnecessary. The question references a specific text and asks for incident specific information, which will eliminate all other opinions/publications from consideration. When provided a reading list, be sure to purchase, read and study those books.

Fire Scenario #8 - Chimney Fire

Photo By: Rich Schwarzenberg

You are a newly appointed tour commander. It's 1:45PM on a brisk Monday afternoon in November. You receive a call from a resident at 12 Oakwood Avenue, who is reporting an odor of smoke outside of his home. As you approach the street, you see what appear to be flames coming from the top of a chimney of a two story, wood-frame single family dwelling at 101 Main Street, which is 2 blocks away and downwind from the caller's home. You keep all units responding to the initial caller's home, but approach 101 Main Street to investigate. Upon arrival, you immediately conduct an exterior size-up and recognize signs of a chimney fire, such as a rumbling noise that resembles a low flying airplane; flames, sparks, and dense smoke extending from the top of the chimney; and products of combustion emanating from existing cracks in the chimney. There is no sign of anybody in the area.

Your department has just purchased its first and only chimney kit and placed it on Engine 1. The kit consists of a mirror, heat gloves, a weighted chimney chain, and chimney bomb (zipper-seal bags containing dry-chemical powder), which are all placed inside a metal bucket.

Scenario Question

As the IC, what actions would you take to mitigate this incident and how would you use the chimney kit?

If you did not have a chimney kit, how would you extinguish a chimney fire?

61

Question #1: What actions would you take to mitigate this incident and how would you use the chimney kit?

1. **En route** – RPM

   - ☐ Monitor <u>R</u>adio Reports

   - ☐ Review <u>P</u>re-incident plans

   - ☐ <u>M</u>ulti-sided view

2. **Upon Arrival** – Position IC CAR PPE

   - ☐ Position vehicle properly

   - ☐ Establish Command

   - ☐ Determine Conditions, Actions, Resources (CAR)

   - ☐ Ensure all members are wearing appropriate PPE and respiratory protection

   - ☐ Re-Direct E1 and L1 to your location. Keep E2 and the BC responding to the initial call, remember the chimney fire is downwind of Oakwood Avenue.

   - ☐ Continue to monitor radio reports

3. **Initial Radio Report** – FM Radio

   - ☐ Establish Tactical Radio <u>F</u>requency (Two potential incidents happening simultaneously)

   - ☐ Determine/Announce Operational <u>M</u>ode: (Offensive)

   - ☐ IRR: Floors and construction type, Fire conditions, size up concerns, name and Location of the CP (Main Street Command, A side of the building)

4. **Size-Up** – COAL TWAS WEALTHS

   - ☐ Apparatus and Staffing: two incidents/investigations are occurring simultaneously.

- ☐ Life Hazard: no one is outside and the fireplace is active.
- ☐ Location and Extent: fire may have extended into the attic, or elsewhere within the home.

5. **Resources** – A RULES WAR+

- ☐ <u>A</u>dditional alarms – 2nd – to ensure adequate personnel at the chimney fire and previous investigation.
- ☐ <u>R</u>apid Intervention Crew (RIC) - for firefighter safety
- ☐ <u>U</u>tility Companies (gas, electric, water) - for utility control
- ☐ <u>L</u>aw enforcement - for traffic/crowd control
- ☐ <u>E</u>MS (BLS/ambulance, ALS/paramedic) - for patient triage and treatment
- ☐ <u>S</u>afety Officer (SO) - for scene safety
- ☐ <u>W</u>ater Supply Officer (WSO) - to ensure adequate water
- ☐ <u>A</u>ccountability Officer - to track and account for all on scene personnel
- ☐ <u>R</u>ehab Unit – for rest and rehabilitation

6. **Strategy / Tactics** - List the objectives and assign companies

Primary objectives include:

- ☐ Locate, confine and extinguish the fire
- ☐ Search for and remove civilians from danger
- ☐ Check for fire extension

Engine Company Ops:

- ☐ Position apparatus past the building for fire attack, leave room in front for the ladder company.
- ☐ Establish a primary (and secondary) water supply.

63

- ☐ Choose/position appropriate size hose lines (Have an engine company stretch a 1 3/4" hose line to the front door as a precaution. Be prepared if the fire extends from the chimney to the structure).

- ☐ Initiate your attack by taking the following actions:

  - ➢ Stop or reduce the flow of oxygen to the flue. This can be as simple as closing the units door(s) and closing any air intakes.

  - ➢ Spread salvage covers in front of the fireplace (a.k.a. the firebox).

  - ➢ Place floor runners (tarps) from the front door to the firebox to keep the floor clean.

  - ➢ Confine and extinguish the fire.

- ☐ Have Engine 1 report to the roof. Use Engine 1's chimney kit (a metal bucket containing a mirror, heat gloves, a weighted chimney chain, and chimney bomb) to take the following actions:

  Interior sector will:

  - ➢ Advance a hose line to the front door.

  - ➢ Send a company to the attic to make sure the fire has not extended into the attic/cockloft.

  - ➢ Place the wood and ashes from the firebox in a fire safe salvage bucket, bring it outside, unload the contents, and hose it down.

  Roof sector (Engine 1 personnel) will:

  - ➢ Cautiously remove the chimney cap, bird screens, or spark arrestors with a hand tool.

  - ➢ Inspect the chimney using the mirror.

  - ➢ If there is fire, drop the chimney bombs down. When they reach the firebox, the bag will burst and the normal draft will carry the powder up to extinguish the creosote.

  - ➢ Slowly lower the weighted chimney chain from the top of the flue to the firebox. Spin the chain to knock the creosote from the walls onto the firebox where it can be extinguished with water or a dry-chemical extinguisher.

  - ➢ Locate, Confine, & Extinguish the Fire.

- ☐ Protect life hazards.

- Protect any threatened exposures with charged hose lines.

- Utilize TIC's to assist with searches.

- Check for extension, check the attic.

- If you suspect fire extension in multiple areas, stretch additional lines.

- Coordinate with ladder companies (and other companies on scene).

- Provide periodic progress reports to the IC.

Ladder Company Ops: LOVERS-UPS (TIC-COP)

- <u>P</u>osition apparatus in front of building for aerial use – extend to the roof for access to chimney.

- Raise and position <u>L</u>adders – raise ground ladder to second floor for secondary egress.

- Force <u>E</u>ntry wherever needed to access all areas of structure.

- Primary and Secondary <u>S</u>earches of all floors (buildings) must be performed.

- <u>R</u>escue Operations are priority #1.

- <u>V</u>entilation as needed – if the interior is filled with smoke, perform horizontal ventilation using a positive pressure fan.

- <u>U</u>tility control (Shut the electric to the fire building if necessary).

- <u>S</u>alvage techniques – use salvage covers in front of the fire place, and as a runner from the door to cover carpets/flooring.

- <u>O</u>verhaul thoroughly to ensure no fire extension- place fire retardant salvage covers in front of the fireplace and limit the number of personnel walking through that area. If the owner/occupant has a vacuum or broom, clean around the unit.

- Utilize <u>TIC's</u> throughout the operation.

- <u>C</u>oordinate with engine companies (and other companies on scene).

- Provide periodic <u>P</u>rogress reports.

Additional tactical tips for personnel:

- Consistently monitor oxygen levels for carbon monoxide (CO). Chimney fires can cause the flue to fail and spill CO and other byproducts of combustion into the walls, ceilings, attics and other hidden spaces.

- ☐ All companies should search for extension.

- ☐ Look for discoloration of surface materials, smoke coming from cracks, outlets, lighting fixtures, or roof coverings.

- ☐ Send a recon team to the attic to check for extension.

- ☐ Inspect the firebox and as much of the chimney in the inside of the home as possible.

- ☐ Consider the possibility of failed internal connectors, which may result in a house fire.

Tips for firefighters working on the roof include:

- ☐ Roofs may be pitched and difficult to access. Work off a platform whenever possible.

- ☐ There may be ice and snow on an already dangerous roof.

- ☐ Consider the extra weight; limit the number of firefighters on the roof.

7. **IMS** – U SIL FLOP SR

- ☐ Unified Command Staff (not necessary)

- ☐ Safety officer (to ensure a safe working environment)

- ☐ Information officer (to communicate with the media and concerned citizens)

- ☐ Liaison officer (to communicate w/ other agencies & ensure smooth operations)

- ☐ Operations officer should be assigned (consider first or second arriving BC)

- ☐ Also assign:

  - ▪ Staging officer

  - ▪ Rehabilitation officer

  - ▪ Interior and Roof Divisions and/or Groups

  - ▪ Division supervisors

8. **Benchmarks** – Time, Progress, RER/LIP

- ☐ Time Management (Progress reports every 10-15 min)
- ☐ RER/LIP

9. **Under Control** – A SOS PAR ICS DO

- ☐ <u>A</u>nnounce via radio
- ☐ <u>S</u>econdary Searches
- ☐ <u>S</u>alvage and <u>O</u>verhaul
- ☐ <u>P</u>AR
- ☐ <u>I</u>nvestigation Unit
- ☐ <u>C</u>O levels
- ☐ <u>S</u>ecure the Building
- ☐ <u>D</u>emobilize
- ☐ Turn the building over to the <u>O</u>wner
- ☐ Advise occupants to have chimney inspected by a certified chimney inspector and cleaned before using it again.

10. **Terminating/Transferring Command** – TD DOP

- ☐ <u>T</u>ransfer or Terminate
- ☐ <u>D</u>ebriefing
- ☐ <u>D</u>ocument/reports
- ☐ <u>O</u>ffer CISD
- ☐ <u>P</u>IA

Question #2: How would you extinguish a confirmed chimney fire without a chimney kit?

- ☐ Briefly open the draft stop and completely discharge a dry-chemical extinguisher upward. (Beware. This method will add oxygen to the fire and temporarily accelerate it. It

- [ ] will also be messy, so be sure to put salvage covers and tarps down before doing so).

- [ ] Water extinguishers are an option; however, most professionals don't advocate using water because of the fear that water will rapidly cool the flue and cause permanent damage to masonry and flue liners. When using a water extinguisher to extinguish the remaining contents in the firebox, close the draft. This will reduce the flow of oxygen into the flue and help with complete extinguishment.

- [ ] If the fire extends from the chimney, treat as a structure fire.

Confined Chimney Fires   BONUS!

All fires are unpredictable, even those that seem confined to a chimney. The most common cause of chimney fires is the ignition of creosote in the flue. Similar to furnaces, wood-burning fireplaces and stoves are designed to safely contain fires. Fireplace chimneys are designed to expel byproducts of combustion (smoke, gases, unburned wood particles, and so on). When these substances rise upward by convection into the relatively cooler chimney flue, condensation occurs and a black/brown residue called creosote sticks to the inner walls. Creosote, which can accumulate in many forms (tar-like, drippy, shiny, hard, crusty, or flaky) is highly combustible; when it builds up, a fire can occur. Creosote accumulates with restricted air supply or cooler-than-normal chimney temperatures, which happens more frequently with chimneys outside the home rather than those that run through the center of the house.

There are several factors that can lead to creosote buildup, such as poor ventilation, failure to maintain a proper temperature inside the flue, burning wet wood, or failure to clean the chimney on a regular basis. Regardless of how it occurs, the longer the smoke remains in the flue, the more likely it is that creosote will form and eventually catch fire inside the chimney flue, resulting in a chimney fire.

Chimney fires have distinct characteristics that firefighters should look for on arrival. These include a rumbling or roaring noise that resembles a freight train or low flying airplane; flames, sparks, and dense smoke that extend from the top of the chimney; and products of combustion emanating with velocity from existing cracks in the chimney mortar. If these signs do not exist, you may still have a chimney fire, but a smaller one that is not visible from the exterior. If a fire extends outside of the chimney and ignites combustible materials or makes its way into the walls, it should be treated as a structure fire.

Photo By: Ron Jeffers

It's a Wednesday evening in August. The wind is blowing at a slight 2-3 mph in the southern direction. You respond to a report of a fire on the 9th floor of Main Street Towers –a 24 story, fire-resistive, residential high rise located at 100 Main Street. The initial response to this incident is 3 Engine companies, 1 Ladder company, and one chief officer.

Upon arrival, you are met by several residents who are still in their pajamas. They claim they were awaken by smoke detectors and opened their doors to find a significant amount of smoke on the 9th floor. You know from your pre-plans that this building has three stairwells (on the B, C and D sides), four elevators (in the center), and one utility shaft way. The building also has three standpipe systems, but the one in the D stairwell, which is on the northern side of the building, has been out of service for more than a month. Two of the stairwells (B and D sides) lead to the roof.

Inside the building, three of the four elevators were returned to the lobby. The fourth appears to not be working. Your aerial ladders are 90' long.

Scenario Question

You were the first fire department representative on scene. What actions would you take from arrival to termination of command?

Fire Scenario #9 - Answer Key

1. **En route** – RPM

   ☐ Monitor <u>R</u>adio Reports

   ☐ Review <u>P</u>re-incident plans

   ☐ <u>M</u>ulti-sided view

2. **Upon Arrival** – Position IC CAR PPE

   ☐ Position vehicle properly

   ☐ Establish Command

   ☐ Determine Conditions, Actions, Resources (CAR)

   ☐ Ensure all members are wearing appropriate PPE and respiratory protection

3. **Initial Radio Report** – FM Radio

   ☐ Establish Tactical Radio <u>F</u>requency.

   ☐ Determine/Announce Operational <u>M</u>ode: (Offensive).

   ☐ IRR: Floors and construction type, Fire conditions, size up concerns, name and location of the CP (Main Street Towers Command, located in the lobby of the building)

4. **Size-Up** – COAL TWAS WEALTHS

   ☐ Construction – stairwells/elevators/shafts (fire travel concerns)

   ☐ Apparatus/staffing – going to need additional alarms at confirmed high rise fires

   ☐ Life hazard – high, especially considering night time.

   ☐ Water supply – you will need to establish secondary water supply.

   ☐ Auxiliary appliances - one standpipe is out of service

   ☐ Weather – warm weather, intensifies rehab concern

70

- Location/Extent of fire – bottom 1/3rd of building, looks like apartment fire from the photo

- Time of day – again, major life hazard concern, sleeping occupants.

- Height – always a concern in high rises

- Special considerations – one elevator has not made it to the lobby

5. **Resources** – AA RULES WAR+

- <u>A</u>dditional alarms – 3rd on scene, 4th in staging – to ensure adequate personnel

- AL<u>S</u>O: <u>A</u>dditional chief officers and command staff personnel

- <u>R</u>apid Intervention Crew (RIC) - for firefighter safety

- <u>U</u>tility Companies (gas, electric, water) - for utility control

- <u>L</u>aw enforcement - for traffic/crowd control

- <u>E</u>MS (BLS/ambulance, ALS/paramedic) - for patient triage and treatment

- <u>S</u>afety Officer (SO) - for scene safety

- <u>W</u>ater Supply Officer (WSO) - to ensure adequate water

- <u>A</u>ccountability Officer - to track and account for all on scene personnel

- <u>R</u>ehab Unit – for rest and rehabilitation

- Red Cross - for displaced occupants

- Rescue Company – due to high life hazard (calls for help)

6. **Strategy / Tactics** - List the objectives and assign companies

Primary objectives include:

- Locate/Confine/Extinguish the fire

- Remove civilians from danger

- Ventilate the building

As the IC, you should:

- ☐ Establish liaison with the building manager/engineer.

- ☐ Attempt to verify which floor the fire is on.

- ☐ Have members access and gain control of the buildings systems:

    - HVAC,

    - communications,

    - elevators,

    - fire pumps,

- ☐ Establish an operations post one floor below the fire (8).

- ☐ Establish a staging area (EMS/Rehab) two floors below the fire (7).

- ☐ Designate an attack and stairwell (B or C side).

- ☐ Designate an evacuation stairwell (B or C side).

- ☐ Designate a ventilation stairwell (D side – busted standpipe).

- ☐ Assign a victim tracking coordinator – building manager.

- ☐ Evacuate affected areas .

First Arriving Engine Company

- ✓ Position for attack.

- ✓ Establish Primary Water Supply.

- ✓ Supply the FDC .

- ✓ Control the Elevator.

- ✓ Take the elevator two floors below the fire, connect to standpipe.

- ✓ Initiate Attack with a 2-1/2" hose line.

- ✓ Locate, Confine, & Extinguish the Fire.

- ✓ Protect Life Hazards, Search as you go.

- ✓ Utilize Thermal Imaging Cameras.

- ✓ Coordinate with ladder companies (and other companies on scene).

✓ Provide periodic radio reports.

Second Arriving Engine Company

✓ Position for attack.

✓ Establish a Secondary Water Supply.

✓ Take the elevator two floors below the fire, take the stairwell to the fire floor.

✓ Stretch a 2-1/2" hose line to back up the first, (or go to the floor above).

✓ Protect Life Hazards, Search as you go.

✓ Check for extension.

✓ Utilize Thermal Imaging Cameras.

✓ Coordinate with ladder companies (and other companies on scene).

✓ Provide periodic radio reports.

Third Arriving Engine Company

✓ Rapid Intervention Crew, one floor below the fire.

Ladder Company Ops: LOVERS-UPS (TIC-COP)

☐ Position the apparatus in front of the building – the Aerial Ladder is too short to reach the fire floor, but you may need to use it for rescue on other floors, line advancement, or defensive ops.

☐ Force Entry wherever needed to access all areas of the structure.

☐ Primary and Secondary Searches of fire floor and floors above the fire must be performed.

☐ Rescue Operations are priority #1 - inform occupants of the designated evacuation stairwell, assist if needed – Ensure there is no one trapped in the elevator that appears to be out of service.

☐ Ventilation (Horizontally: by opening windows, Vertically: by using the designated ventilation stairwell on the D side).

☐ Control the building systems and Utilities.

- ☐ Salvage techniques should be used whenever possible.

- ☐ Overhaul thoroughly to ensure there is no fire extension.

- ☐ Utilize TIC's throughout the operation.

- ☐ Coordinate with engine companies (and other companies on scene).

- ☐ Provide periodic Progress reports.

  - ✓ Take the elevator two floors below the fire.

  - ✓ Check for extension as you go (shafts, utility openings, plenum space, etc.).

7. **IMS** – U SIL FLOP SR

- ☐ Unified Command Staff (Have a representative of law enforcement, EMS, and a building manager close by)

- ☐ Safety officer (to ensure a safe working environment)

- ☐ Information officer (to communicate with the media and concerned citizens)

- ☐ Liaison officer (to communicate with other agencies & ensure smooth operations)

- ☐ Operations officer should be assigned (consider first or second arriving BC)

- ☐ Assign a Finance officer/section

- ☐ Assign a Logistics officer/section

- ☐ Assign an Operations officer

- ☐ Assign a Planning officer/section

- ☐ Assign a Staging officer/location

- ☐ Assign a Rehabilitation officer

- ☐ Establish Divisions/Groups

- ☐ Assign division supervisors

8. **Benchmarks** – Time, Progress, RER/LIP

- ☐ Time Management (Progress reports every 10-15 min)

- ☐ RER/LIP

9. **Under Control** – A SOS PAR ICS DO

- [ ] <u>A</u>nnounce via radio

- [ ] <u>S</u>econdary Searches

- [ ] <u>S</u>alvage and <u>O</u>verhaul

- [ ] <u>PAR</u>

- [ ] <u>I</u>nvestigation Unit

- [ ] <u>C</u>O levels

- [ ] <u>S</u>ecure the Building

- [ ] <u>D</u>emobilize

- [ ] Turn the building over to the <u>O</u>wner

10. **Terminating/Transferring Command** – TD DOP

- [ ] <u>T</u>ransfer or Terminate

- [ ] <u>D</u>ebriefing

- [ ] <u>D</u>ocument/reports

- [ ] <u>O</u>ffer CISD

- [ ] <u>P</u>IA

Planning and Practice

BONUS!

High rise firefighting is an area of our business that requires a tremendous amount of education, planning, and training. It is extremely important to use practice scenarios like this one to prepare for the challenges that you will encounter at these types of incidents. For those who have little to no experience in this area, the need to identify and prepare for the challenges that lie ahead becomes even more evident once you begin trying to organize a structured response to the way you would tackle a significant high rise fire. Add trapped occupants to the mix and you can understand why planning and practice is so important.

You are the Incident Commander at a three family residential fire. The fire started in the unfinished basement where one resident stored multiple gasoline cans for a generator. Upon arrival, it is immediately obvious the fire had spread through the balloon frame structure up to the top floor. Early in the operation, on scene personnel stretch two lines to the basement, one line to the first floor and another to the attic and still are not able to make progress on the fire. Due to staffing challenges, only one ladder company is on scene, another two mutual aid companies are en route. The on scene Ladder company personnel are conducting an interior search when it becomes obvious from multiple reports of fire extension that you have to change your tactics from offensive to defensive.

Scenario Question

What actions would you take to change from an offensive to a defensive strategy? Be sure to list signs to consider when determining collapse potential?

1. **Announce a move to defensive operation via radio.**

   ❏ Have your dispatcher announce a change in tactics from offensive to defensive, and order all personnel operating within the structure to evacuate and meet at the command post (or another designated location).

2. **Sound the evacuation tones and air horn.**

   ❏ Have your dispatcher send a designated evacuation tone over the radio.

   ❏ Have apparatus drivers activate their air horn four times to signify evacuation of the structure.

3. **Request additional alarm(s).**

   ❏ Request an additional alarm and necessary resources if you do not have enough on scene or in staging to handle the change in tactics.

4. **Conduct a personnel accountability roll call (PAR).**

   ❏ Along with your accountability and safety officers, account for all personnel to ensure they have made it out of the structure. Have your dispatcher call the officer of each company to confirm all personnel are accounted for. (See the following example)

     - Dispatcher: "Engine 1."

     - Engine 1 officer: "E-1 officer, ALL members are accounted for."

     - Dispatcher: "Engine 2 officer."

     - Continue calling units until all companies/members are accounted for.

   ❏ If members do not respond, activate the rapid intervention crew (RIC).

5. **Readjust your IMS to reflect the new defensive operation.**

   ❏ Establish divisions on all 4 sides (A, B, C, and D) *after* you conduct a PAR and account for all members.

6. **Establish your collapse zone(s).**

- ❏ Walls collapse in three general manners.

  - 90-degree-angle collapse: This is most common and is similar to a falling tree. The wall falls straight out, and the top hits the ground at a distance equal to the height of the wall.

  - Curtain-fall collapse: Generally occurs with a masonry wall. It collapses like a curtain dropping from the top, creating a pile of debris at the base of the wall.

  - Inward/outward collapse: A wall leaning inward may not necessarily fall inward. The lower or upper portion may slide or "kick" outward.

- ❏ The collapse zone itself should be as wide as the structure and one and a half times the height.

- ❏ Take construction materials into consideration.

  - Ordinary and heavy timer buildings = two times the height of the structure.

- ❏ Use caution/barricade tape to clearly mark the edges of the collapse zone.

- ❏ When established, collapse zones must be maintained during and after the incident, during the investigation, and until the structure is examined by an engineer.

- ❏ Assign additional safety officer(s) to cover all four sides.

- ❏ If you haven't already, call for the response of the utility companies to shut off the gas, electric, and water to the building from an exterior locations, away from structure.

... Continue on page 80

7. **Monitor for signs of collapse.**

   ❏ Depending upon the height of the structure and its building features, set up a number of surveyors/transits to detect an early structural movement from walls, church steeples and bell towers, water tanks, etc.

   ❏ Consider the following when determining collapse potential:

☐ Fire size and location	☐ Spongy roof or floors
☐ Heavy fire for an extended period of time	☐ Previous fires at this location
☐ Pieces of the building falling off	☐ Explosions, flashovers, or backdrafts
☐ Cracks in walls	☐ Water and/or smoke pushing through solid masonry wall
☐ Leaning or bowing walls	
☐ Building age and condition	☐ Smoke through mortar joints
☐ Faulty/poor construction	☐ Accidental cutting of structural support members
☐ Foundation failure	☐ Lightweight construction components
☐ Extraordinary loads	☐ Extreme weather conditions
☐ Lack of water runoff	☐ Fire reaches the truss roof
☐ Sagging floors or beams	☐ Unusual noises (creaking)

   ❏ Be mindful of the condition of the parapet, canopy, marquee, cornice, floors, and roof.

   ❏ Constantly monitor for secondary collapse from the existing structure or collapse and failure of any surrounding exposure buildings.

   ❏ Note: If collapse occurs, follow the Non-Fire Scenario format for Structural Collapse.

8. **Set up master streams (ground monitors, deluge guns, large-diameter hose lines, and so on).**

   ❏ If the roof is still intact, aim the streams up, through windows, into involved ceilings.

   ❏ If the roof has burned away, use elevated streams and aim down into the building.

   ❏ If possible, position and secure unmanned master streams outside the collapse zone.

   ❏ If offensive operations are occurring in another portion of a large building (for example, a firefighter rescue is in progress) make sure master streams are not flowing in that area.

9. **Secure an additional water supply (from another source or water main).**

   ❑ Whenever possible, do not use the same water main when additional water is needed.

10. **Protect nearby exposures.**

    ❑ Do everything in your power to protect exposures from collapse, radiant heat, water runoff, etc.

    ❑ Evacuate exposures, if necessary.

11. **Assign a brand patrol. (This will depend on the buildings contents, height, and construction.)**

    ❑ Use a minimum of one engine and ladder company.

    ❑ Position downwind to track flying brands.

12. **Rotate personnel frequently.**

    ❑ Establish emergency incident rehabilitation

Signs Dictate Procedures

BONUS!

When reviewing scenario narratives, it's easy to become so focused on tactics that you overlook a word or sentence that was strategically placed there to lead you in another direction. Know and understand the signs that lead to conditions like collapse, backdraft, flashover, rapid fire travel, and other conditions that would dictate a swift change in procedures, so you can address them properly.

Non-Fire Scenarios

It is hard to identify and list every possible incident, but in this section, I will provide you with tips on non-fire scenarios that include hazardous material incidents, structural collapse, water main breaks, overturned tractor trailers, school and public transportation bus accidents, compounded incidents, and more. These are the more challenging types of incidents you may encounter in both the field and the assessment center.

At this point in the book, you should be well underway with the direction and concepts of how you can begin preparing for scenarios. Whether you're being evaluated in a promotional exercise or coordinating operations in the street, the following non-fire incidents are designed to assist you in further preparing for your goals of becoming a seasoned fire officer.

Photo By: Brett Dzadik

Non-Fire Scenario #1
Hazardous Material Incidents

Hazardous material incidents are not unique to any one jurisdiction, town, city, or state. They are universal to us all. They can happen anytime, anywhere, and to anyone. From the thousands of chemical types and uses, to their modes of transportation and the challenges that come from their involvement, firefighters need to be well-educated and trained to deal with the unknown.

Because of this, fire officers are expected to know and understand the necessary actions to take when responding to hazardous material incidents. The following guideline provides you with a simple but comprehensive format to follow when responding to incidents and questions about these types of incidents. You will notice that this format begins and ends the same way a fire scenario does. The main difference lies in the strategy and tactics area, where you should implement some form of hazmat response guideline. In this instance, I refer to Hildebrand and Noll's 8-step hazardous materials guideline which is widely used within the fire service.

The Haz-Mat Format

1. **En route** – WARP

   There are a few things you can and should do on your way to the scene: Review <u>P</u>re-incident plans or tactical survey information specific to the area/occupancy or site, Monitor <u>R</u>adio Reports, <u>A</u>pproach the incident area from Uphill and Upwind, and request <u>W</u>ind speed and direction and projected weather from the dispatch center.

2. **Upon Arrival** – Position ICU CAR PPE - SUB

   Establish Command if you are the first to arrive. Prepare for a <u>U</u>nified command. If you are relieving a current Incident Commander (<u>IC</u>), have a brief face-to-face meeting with that individual for a quick incident briefing where the following information is relayed: Conditions upon arrival, Actions already taken, Resources already ordered (<u>CAR</u>). In other words, find out answers to the following three questions: What do you have? What did you do? *and* What do you need? When relieving another IC, be sure to *Assume* Command and assign that individual to another position (such as operations or interior division chief). You should also be conscious of where you <u>Position</u> your vehicle. Inform in-coming units of a Safe approach, Uphill and upwind. Use Binoculars to size-up the incident. Ensure that all members who work on scene wear appropriate PPE throughout the duration of incident.

3. **Initial Radio Report** – FM Radio

Establish your tactical radio Frequency. Determine your operational Mode (Offensive, Defensive, Combination, or Non-intervention). Then begin transmitting your Initial Radio Report (IRR), which should consist of: Conditions, number of known victims, obvious size up concerns, and the name and location of the command post.

Note: Remember to request periodic progress reports throughout the incident.

4. **Size-Up** – COAL TWAS WEALTHS

Conduct a thorough 15 point size-up and address any conditions that require your attention. The 15 points are: Construction, Occupancy, Apparatus/staffing, Life hazard, Terrain, Water supply, Auxiliary appliances, Street conditions, Weather, Exposures, Area, Location/extent of fire, Time of day, Height and Special considerations.

5. **Resources** – A RULES WAR HOG+

There are resources you should call for at every incident, regardless of type, size and scope. The acronym above is one way to remember them. These are not the only resources you will call (or assign) at a hazardous material incident, but rather the minimum list of the essential ones.

☐ Additional alarms - to ensure an adequate number of personnel and resources.

☐ Rapid Intervention Crew (RIC) - for firefighter safety.

☐ Utility Companies (gas, electric, water) – possibly, for utility control.

☐ Law enforcement - for traffic/scene/crowd control.

☐ EMS (BLS/ambulance, ALS/paramedic) - for patient triage and treatment.

☐ Safety Officer (SO) - for scene safety.

☐ Water Supply Officer (WSO) – possibly, to ensure adequate water.

☐ Accountability Officer - to track and account for all on scene personnel.

- ☐ U̲nderline Rehab Unit – for rest and rehabilitation .

- ☐ H̲azardous Material Team.

- ☐ O̲ffice of Emergency Management (OEM),

- ☐ G̲overnment agencies (city, state, federal) such as sewage, water treatment, Coast Guard, Environmental Protection Agency, etc.

The plus (+) represents all additional resources that will be specific to the scenario. Some examples include: Red Cross (for displaced occupants), Dept. Public Works (for sand and salt), Health department, USAR, Rescue Co, Air supply/mask service unit for incidents that may be potentially long in duration, High-volume foam unit, EMS mass-casualty units, the need to notify area hospitals, etc.

6. Strategy / Tactics – SHIP IRDT

The best incident management system (IMS) in the world is worthless without sound objectives, strategy and tactics. This is where a well prepared and intelligent officer will shine. The most effective way to organize your approach is by using a format such as the 8 Step Plan for tactical management of hazmat incidents. Those steps are: SHIP IRDT

1. Site Management and Control

- ☐ Establish a security perimeter by isolating and denying entry to the area/building. Give significant consideration in using the police to assist you with this responsibility.

- ☐ Establish control zones for the incident site: Hot, Warm, and Cold.

- ☐ Assure a safe approach for incoming resources.

- ☐ Establish staging (Uphill and Upwind) as a method of controlling arriving resources.

- ☐ Implement public protective actions by evacuating, protecting in place, or implementing a combination of the two.

2. Identify the Problem

- ☐ Determine building occupancy and location; the name on the front of the building may tip you off to specific concerns.

- ☐ Container shape can indicate pressurized vs. non-pressurized containers.

☐ Marking and colors as well as placards and labels can provide product information and mitigation options for first responders.

☐ Shipping papers, facility documents and Material Safety Data Sheets (MSDS) will provide product information.

☐ Monitoring and detection equipment can identify the presence of a product.

☐ Your senses can help identify the problem. Pay attention to what you see, hear, and smell.

3. <u>**H**azard & Risk Evaluation</u>

Hazard: this is defined as the Physical and Chemical Properties of a material.

Risk: the intangible, undesirable probability of suffering harm or loss.

Gather hazard data from the following:

☐ Reference materials (Department of Transportation Emergency Response Guidebook – DOT/ERG).

☐ Technical information centers such as CHEM-TREC.

☐ Hazardous material databases.

☐ Right-to-know information; Material Safety Data Sheets (MSDS).

☐ Monitoring instrument.

☐ Determine the extent of damage to container.

☐ Predict the likely behavior of the released material and containers.

☐ Analyze HAZARDS & RISKS to determine the safest and most effective Incident Action Plan.

4. <u>Select proper **P**PE & Protective Clothing</u>

☐ Is HazMat gear/equipment needed, or will firefighting gear/respiratory protection be enough?

5. <u>Information Management and Resource Coordination (ICS)</u> - U SIL FLOP SR

Use the Incident Command System at ALL haz-mat incidents, regardless of size/scope. Assignments may include the following command and support staff members:

- ☐ Unified Command Staff

- ☐ Safety officer (to ensure a safe working environment)

- ☐ Information officer (to communicate with the media and concerned citizens)

- ☐ Liaison officer (to communicate with other agencies and ensure smooth operations)

- ☐ Finance officer/section

- ☐ Logistics officer/section

- ☐ Operations officer

- ☐ Planning officer/section

- ☐ Staging officer

- ☐ Rehabilitation officer

Also: Establish Divisions/Groups early to enhance communications and improve accountability. Don't wait to be overwhelmed with your span of control. (Example: Rescue = Rescue Group; Spill containment = Containment Group, etc.) Assigning division supervisors will help you manage your span of control. Less people trying to communicate with you directly ensures a greater level of effectiveness.

6. Implement **R**esponse Objectives

The level of risk will determine how and if you deploy your hazardous material and mitigation team(s). From the hazard and risk evaluation, start off with determining:

- ☐ Offensive operations—Rescue, containment, etc

- ☐ Defensive operations—confinement

- ☐ Combination of both, or

- ☐ Non-intervention—it takes its natural course

From here you can identify more specific actions and objectives from the challenges presented.

- ☐ Rescue—for example, extricate the driver.

- ☐ Spill control/confinement—for example, dike and dam the spill.

- [ ] Leak control/containment—for example, plug, patch, shut valves.

- [ ] Fire control—for example, eliminate ignition sources, de-energize and stabilize the vehicle, suppress vapors with foam, extinguish the fire.

- [ ] Public protective actions—for example, evacuate or protect in place the occupants of a nursing home or day care center.

*Assign engine and ladder companies (see below), especially if there is a fire at the incident:

Engine Company Ops: Here is a list of the key points you will want to address with engine companies.

- [ ] Position apparatus

- [ ] Establish a primary (and secondary) water supply

- [ ] Initiate attack

- [ ] Choose appropriate size hose lines

- [ ] Advance and position hose lines

- [ ] Locate, Confine, & Extinguish (LCE) the fire

- [ ] Protect life hazards

- [ ] Protect exposures

- [ ] Supply auxiliary appliances

- [ ] Utilize Thermal Imaging Cameras (TIC)

- [ ] Coordinate with ladder companies (and other companies on scene)

- [ ] Provide periodic progress reports

Ladder Company Ops: The goals of all ladder companies at most incidents can be remembered by using the acronym LOVERS-UPS (TIC-COP). Note: That acronym, underlined below, is not in order.

- [ ] Position apparatus

- [ ] Raise and position Ladders

- ☐ Force Entry
- ☐ Primary and Secondary Search
- ☐ Rescue Operations
- ☐ Ventilation (Horizontal, Vertical, Outside Vent Member)
- ☐ Utility control
- ☐ Salvage
- ☐ Overhaul
- ☐ Utilize TIC's
- ☐ Coordinate with engine companies (and other companies on scene)
- ☐ Provide periodic Progress reports

7. Decontamination

- ☐ Assume anything coming out of the "HOT ZONE" is exposed and potentially contaminated.
- ☐ Establish a decon site/officer and Group.
- ☐ Conduct DECON and medical monitoring for all members.

8. Terminate the incident (see #9 for steps to take)

7. **Benchmarks** – Time, Progress, RER/LIP

Time management helps the IC measure if tactics are working. Review, Evaluate, and Revise your strategy and tactics periodically and use progress reports to ensure you are meeting your objectives of Life safety, Incident Stabilization, and Property Conservation (RER/LIP).

- ☐ Time management: Note the time of incident start – request Progress reports every 10-15 min
- ☐ RER/LIP

8. **Under Control** – A SOS PAR ICS DO

When the incident is under control, <u>A</u>nnounce via radio then conduct the following actions:

- ☐ Complete <u>S</u>econdary Searches.

- ☐ Meet <u>S</u>alvage and <u>O</u>verhaul responsibilities.

- ☐ Conduct a Personnel Accountability Roll Call (<u>PAR</u>) to account for personnel.

- ☐ Request an <u>I</u>nvestigation Unit for cause and determination.

- ☐ Check <u>C</u>O levels.

- ☐ <u>S</u>ecure the building / area.

- ☐ <u>D</u>emobilize the Incident.

- ☐ Turn the building / area over to the <u>O</u>wner.

9. **<u>Terminating/Transferring Command</u>** – TD DOP

- ☐ <u>T</u>ransfer to another officer or Terminate completely.

- ☐ Conduct an incident <u>D</u>ebriefing.

- ☐ <u>D</u>ocument the incident, complete reports.

- ☐ <u>O</u>ffer Critical Incident Stress Debriefing (CISD).

- ☐ Schedule a <u>P</u>ost Incident Analysis (PIA).

The Haz-Mat Format - Short Version

Once you understand the haz-mat format above, you will want to condense each of the 9 categories down to the least amount of words possible. This will enable you to create a tactical worksheet for real world incidents, as well as practice and promotional scenarios. Here is an example of what it might look like.

1. **<u>En route</u>** – WARP

- ☐ <u>W</u>ind speed/direction and projected weather

- ☐ <u>A</u>pproach Uphill and Upwind

- [ ] Monitor Radio Reports

- [ ] Review Pre-incident plans

2. **Upon Arrival** – Position ICU CAR PPE - SUB

   - [ ] Prepare for a Unified command.

   - [ ] Establish Command (IC), or

   - [ ] Face-to-face with current IC

   - [ ] Incident Debriefing - Conditions, Actions, Resources (CAR)

   - [ ] Assume Command

   - [ ] Re-assign IC

   - [ ] Ensure all members wear PPE and respiratory protection

   - [ ] Inform incoming units of a Safe approach.

   - [ ] Position vehicle Upwind (uphill)

   - [ ] Use Binoculars to size-up the incident.

3. **Initial Radio Report** – FM Radio

   - [ ] Establish Tactical Radio Frequency.

   - [ ] Determine/Announce Operational Mode: (O/D/C/N).

   - [ ] IRR: Conditions, victims, concerns, size up concerns, name and location of the CP.

4. **Size-Up** – COAL TWAS WEALTHS

   - [ ] Construction, Occupancy, Apparatus/staffing, Life hazard, Terrain, Water supply, Auxiliary appliances, Street conditions, Weather, Exposures, Area, Location and extent of fire, Time of day, Height and Special considerations.

5. **Resources** – A RULES WAR HOG+

   - [ ] Additional alarms - to ensure an adequate amount of personnel

- ☐ Law enforcement - for traffic/scene/crowd control
- ☐ EMS (BLS/ambulance, ALS/paramedic) - for patient triage and treatment
- ☐ Safety Officer (SO) - for scene safety
- ☐ Water Supply Officer (WSO) – possibly, to ensure adequate water
- ☐ Accountability Officer - to track and account for all on scene personnel
- ☐ Rehab Unit – for rest and rehabilitation
- ☐ Hazardous Material Team
- ☐ Office of Emergency Management (OEM),
- ☐ Government agencies (city, state, federal) such as sewage, water treatment, Coast Guard, Environmental Protection Agency, etc.
- ☐ (±) Red Cross (for displaced occupants), Dept. Public Works (for sand and salt), Health department, USAR, Rescue Co, Air supply/mask service unit, High-volume foam unit, EMS mass-casualty units, notify area hospitals, etc.

6. **Strategy / Tactics** – Implement 8 Step Plan for (SHIP IRDT)

   1. Site Management and Control
   2. Identify the Problem
   3. Hazard & Risk Evaluation
   4. Select proper PPE & Protective Clothing
   5. Information Management and Resource Coordination (ICS)
   6. Implement Response Objectives
   7. Decontamination
   8. Terminate the incident

7. **Benchmarks** – Time, Progress, RER/LIP

   - ☐ Time Management (Progress reports every 10-15 min)
   - ☐ RER/LIP

    ☐ Rapid Intervention Crew (RIC) - for firefighter safety

    ☐ Utility Companies (gas, electric, water) – possibly, for utility control

8. **Under Control** – A SOS PAR ICS DO

    ☐ Announce via radio

    ☐ Secondary Searches

    ☐ Salvage and Overhaul responsibilities

    ☐ PAR

    ☐ Investigation Unit

    ☐ CO/Air quality levels

    ☐ Secure the Building/Area

    ☐ Demobilize

    ☐ Turn the building over to the Owner or on scene authority

9. **Terminating/Transferring Command** – TD DOP

    ☐ Transfer or Terminate

    ☐ Debriefing

    ☐ Document/reports

    ☐ Offer CISD

    ☐ PIA

Master the 8 Steps

BONUS!

The key to success at Hazardous Material incidents is to approach the scenario with a simple format that will enable you to cover all the necessary steps needed in order to successfully mitigate the incident. The 8 steps is just one optional way to prepare, but a successful one.

Photo By: Ron Jeffers

You are in command at a two alarm structure fire in a two-story, wood frame, multiple dwelling. The fire started in the first floor kitchen and traveled to the third floor via interior voids and exterior siding. The fire was thought to be contained and under control when conditions began to rapidly change for the worse. Ladder 1 personnel are on the roof, on side C of the structure when a rear portion of the roof collapses. They managed to escape; however, there were several firefighters inside the structure at the time of the collapse.

The structure is two houses away from a busy train station. It's 5:45 PM and people who are exiting the train are drawn towards the attention of the incident.

Scenario Question

As the IC, What actions would you take to ensure the safety of your members?

Include detailed information on how you would direct operations if firefighters were hurt, trapped or missing.

1. **Immediately send an URGENT message announcing the collapse**

2. **Immediately conduct a personnel accountability Roll Call (PAR)**

   ☐ If any personnel do not answer the roll call, activate the RIC

   ☐ Use assistance from your Accountability officer and command board

3. **Call for 'additional' resources** – REAL MOS

   ☐ <u>R</u>escue Company

   ☐ Additional <u>E</u>MS personnel

   ☐ Additional <u>A</u>larms (at least one to scene and one to staging)

   ☐ <u>L</u>ighting for extended operations

   ☐ <u>M</u>ask service unit/Air supply

   ☐ <u>O</u>ffice of Emergency Management (OEM) - to help acquire heavy machinery for lifting debris

   ☐ <u>S</u>tructural Engineer

4. **Conduct a quick collapse size-up, consisting of the following**

   The goal is to determine how many (if any) firefighters are hurt, trapped or unaccounted for and survey the scene for safety. To do this, take the following actions:

   ☐ Construction: ID any construction features that could promote further/secondary collapse concerns.

   ☐ Street: Big rigs on a nearby roadway and active trains operating on the nearby rail lines must be halted for the duration of the operation.

   ☐ Weather: Call for projected weather report because it may have an impact on the operation and overall safety of the incident.

   ☐ Exposures: ID any surrounding or attached properties that might be affected by the collapse.

☐ Area: Conduct a 360-degree view of the incident. Determine if the collapse is extensive or localized.

☐ Life Hazard: Determine if any civilians are unaccounted for.

  ➢ Perform a risk analysis: are you confronted with a rescue or a recovery.

☐ Exposures: Determine the collapse type:

  ➢ Lean-to collapse

  ➢ Pancake collapse

  ➢ Unsupported collapse

  ➢ V-type collapse

  ➢ Inward/outward collapse

  ➢ 90-degree collapse

5. **Preparing for collapse rescue operations would include the following:**

☐ <u>Site management and control:</u>

  ☐ Establish a security perimeter by isolating and denying entry to the area/building. Have the police to assist.

  ☐ Remove all non-essential personnel and control the scene.

  ☐ Evacuate all affected areas and any attached or nearby structures.

  ☐ Conduct a site survey. Gather information specific to risk analysis, the location of trapped individuals, what might have caused the collapse, secondary collapse concerns, and accessibility.

  ☐ Compile all information that will assist with the rescue operation and safety of the incident.

  ☐ Establish collapse and operational zones (at least 1.5 times the height of the building).

  ☐ Utility control: eliminate explosive atmosphere, electrocution, and potential drowning risks.

  ☐ Stretch protective hose lines.

- ☐ Shore unstable areas.

- ☐ Conduct air monitoring.

- ☐ Use surveyors transit to detect early structural movement/secondary collapse

- ☐ Eliminate vibrations.

- ☐ Increase supervision in difficult and dangerous areas.

- ☐ Increase lighting in area.

- ☐ Conduct frequent PARS.

- ☐ Conduct frequent reliefs.

- ☐ Provide food and shelter.

6. <u>Conduct the actual rescue operations</u>

- ☐ 1. <u>Surface victim search and removal</u>. If it is difficult to see the top of the collapse pile, place at least one tower ladder over the collapse site to assist with victim identification, search and removal, as well as additional reconnaissance of the site.

- ☐ 2. <u>Void search rescue</u>. When buildings collapse, they create voids. Voids will be created by the type of collapse, as well as the buildings internal stock. Void search rescue is an extensive search of all possible areas in the collapse site where a person may be trapped and still alive. Void search must be done by trained rescue personnel.

- ☐ 3. <u>Selected debris removal and tunneling</u>. As in the void search, this phase of the operation will continue as long as you are still considering this a rescue and not a recovery. Debris removal and tunneling is the most dangerous part of a collapse operation. It is here where rescuers will be removing debris and tunneling into the pile to search for trapped occupants. Removing and tunneling through debris increases the possibility of secondary collapse; only trained members should perform these tasks. This part of your operation can continue for days or weeks.

- ☐ 4. <u>General debris removal and search</u> - You should only begin general debris removal when you are certain that there are no other survivors. (changes from a rescue operation to a recovery operation) This phase of the operation would require additional resources, such as:

  - ➤ Outside agencies with heavy construction equipment/operators will be directed to remove sections of debris and place them in a designated area for further search.

> ➤ Specially trained cadaver dogs can help pinpoint specific debris removal.

> ➤ Scene/critical incident stress debriefing (CISD): the debriefing can be done on-site with a more formal debriefing after the incident or operational period.

7. **Incident scene management will expand to include the following:**

☐ Firefighting efforts should use a second tactical radio frequency. Rescue efforts will continue on the initial frequency (to ensure communication with trapped FF's).

☐ Additional <u>S</u>afety Officer(s)

☐ Liaison officer to coordinate assisting and cooperating agencies

☐ Information officer to inform the media and public

☐ Intelligence officer

☐ Operations officer

☐ Assign Divisions/Group with specific task assignments

☐ Planning and Logistic sections if incident is prolonged.

☐ Establish a staging area and officer for all responding resources.

8. **<u>Benchmarks</u>** – Time, Progress, RER/LIP

☐ Time Management (Progress reports every 10-15 min)

☐ RER/LIP

9. **<u>Under Control</u>** – A SOS PAR ICS DO

☐ <u>A</u>nnounce via radio

☐ <u>S</u>econdary Searches

☐ <u>S</u>alvage and <u>O</u>verhaul

☐ <u>PAR</u>

- ☐ I̲nvestigation Unit
- ☐ C̲O levels
- ☐ S̲ecure the Building (if possible)
- ☐ D̲emobilize
- ☐ Turn the building over to the O̲wner (or the appropriate agency)

10. **Terminating/Transferring Command** – TD DOP M̲

- ☐ T̲ransfer or Terminate
- ☐ D̲ebriefing
- ☐ D̲ocument/reports
- ☐ O̲ffer CISD
- ☐ P̲IA
- ☐ M̲edical follow-ups: to assess cumulative effects of exposure to the unknown

Collapse Rescue and Recovery   BONUS!

Structural collapse incidents are not unique to any one jurisdiction, town, city, or state; they are universal to us all. They can happen anytime, anywhere, and to anyone. From the various types of building construction and structural alterations to the effects from earthquakes, weather, and terrorism, firefighters need to be well educated and trained to deal with the unknown. As with the previous hazardous materials scenario format, the collapse scenario outline above is designed to help you organize and direct your areas of responsibility, and provide a thorough, well thought out, answer. Continue to educate yourself on the areas of collapse rescue listed in this scenario. You can find information about those four specific areas in a variety of publications or credible online websites.

1. Surface victim search and removal

2. Void search rescue

3. Selected debris removal and tunneling

4. General debris removal

Non-Fire Scenario #3
Water Main Break

Photo By: Ron Jeffers

On March 3rd, you respond to a water main break on a main road that runs through your community. The break occurred in a busy residential and commercial area at approximately 10:00 in the morning. Six months earlier your local water department had begun replacing some of the old, underground piping that comprises your community's water distribution system. It had been determined that this system may be prone to failure.

Scenario Questions

1. As the fire officer in charge of the scene, what actions would you take to ensure the safety of your members and the public?

2. What resources would you request?

3. How do you identify the source of a water leak?

Non-Fire Scenario #3 - Answer Key

Question #1:

What actions would you take to ensure the safety of your members and the public?

1. **Approach the scene with caution and park at a safe distance**

   - ☐ Approach slowly and look for possible collapsed roadways resultant from the ground washing away underneath the pavement.

   - ☐ Beware of manhole covers/lids that may have become dislodged; or open manholes that may be hidden under the water.

   - ☐ Exit the apparatus and walk cautiously toward the affected area.

   - ☐ Do not walk blindly into water puddles, pools, or flooded areas.

   - ☐ If you must walk in flooded areas, use a tool (such as a pike pole) to test for solid footing.

   - ☐ Collapse may be imminent, even to firefighters who are out of their vehicles.

   - ☐ Conduct a thorough size-up as you survey the scene.

   - ☐ Wear reflective clothing/vests anytime you are working on a roadway, even if it's closed.

2. **Establish a safety zone:**

   - ☐ Secure the area and ensure that firefighters and civilians remain at a safe distance.

   - ☐ Use caution tape to set up your safety zone and secure the area to keep civilians away.

   - ☐ Shut the road and have the police set up detour signs. Do not let people walk or cars drive over the area.

3. **Establish a unified command:**

   - ☐ Include a representative from your local water authority and law enforcement.

4. **Search for civilians who may be in danger:**

   ☐ Access the structures near the area. Conduct an extensive primary search of all areas affected by the water main break, including exposures (inside and out).

   ☐ Search large puddles, collapsed roadways, flooded areas, under bridges, inside voids, nearby vehicles, nearby basements that may be flooded, and any other areas within the danger zone for civilians who may need assistance, and remove them from danger.

   ☐ Direct ambulatory civilians to safe areas.

   ☐ Initiate rescue efforts if needed.

5. **Conduct periodic checks of exposures:**

   ☐ Check nearby buildings/basements for any developing hazards, such as rising water lines threatening utilities.

   ☐ Shut down utilities if necessary.

6. **Do not leave the scene until the danger has been removed.**

   ☐ Once the water flow stops and the incident is stabilized, transfer command to law enforcement or the water authority as they remain on scene and focus on repair and cleanup.

Question #2:

What resources would you request at a water main break?

   ☐ Additional alarms, if additional personnel are needed for operations or command staffing.

   ☐ Request the utility company to investigate for endangered gas lines or underground electrical lines.

   ☐ Request law enforcement for pedestrian and traffic control.

   ☐ Contact the Red Cross for displaced occupants.

   ☐ In cold weather, when ice is on the roadway, call a municipal agency and request a salt/sand spreader to help avoid slipping/sliding hazards.

Question #3:

How do you identify the source of a water leak?

- ☐ Water coming from the street is an indication of a water main break.

- ☐ If the flow is coming from a sewer grating, the sewer may be backing up. Notify your local department of public works or street department. This will be their responsibility.

- ☐ If water is flowing from a building onto the sidewalk and/or street, this normally indicates a problem from within. Firefighters should enter the structure and attempt to stem the flow by shutting the water supply to the building to avoid further damage. Advise owners to contact a licensed plumber to repair the problem.

Bonus tip for water main breaks:

- ☐ Ask the water authority to provide you with a list of hydrants that will be out of service until the main is repaired.

- ☐ Ask them to notify the fire department when the hydrants are back in service.

- ☐ Ask community officials to notify you when the road is reopened.

Extra Credit
BONUS!

Aging grids and careless excavators are the two biggest reasons for water main failure, but even the simple act of firefighters closing a hydrant, if done too quickly, can create a water hammer that can cause a water main to fail. The broken water main may become immediately noticeable, but it may also begin to secretly wash away the soil under the concrete and asphalt without tipping anyone off that a break in the system exists and/or the fact that serious damage is occurring. The result can be the eventual collapse of a crowded sidewalk or busy street, and there's no way to predict when this failure will occur.

The responsibility of firefighters who respond to water-related incidents is to ensure safety and prevent damage. At a water main break, we can accomplish these two goals by establishing command, identifying the source of the leak, calling in the proper resources, searching all nearby exposures for civilians in danger, establishing safety zones, and helping to stabilize the incident by assisting the water authorities (water department), who will be responsible for shutting down the main and making the necessary repairs.

It's just after 2:45 PM on Memorial Day and you are completing reports from a previous alarm when another alarm is transmitted. This one is for an overturned tanker truck on Route 10 – a four lane highway that runs through your community. Several calls have come in from individuals who are stuck in traffic behind the truck. One caller reports that several people are trying to free the driver, who is trapped in the vehicle.

You respond to the call along with two engines and one ladder company. As you approach the scene, you realize the truck is ½ mile west of a grammar school and ¼ mile east from a busy shopping complex consisting of popular outlet stores.

When you arrive first on the scene, you see a great number of people out of their vehicles, observing and recording the attempted rescue with their cell phones. You also see a large amount of liquid coming from the damaged container that is flowing slowly downhill towards the vehicles, and a hazy cloud is moving downwind. One placard on the side of the truck says flammable liquid. You also discover the accident occurred just prior to an upwind onramp.

The wind is coming from the west at 6 MPH and the temperature is 92 degrees.

Scenario Question

As the IC, what actions would you take at this incident?

1. **En route** – WARP

   ☐ **W**ind speed/direction and projected weather

   ☐ **A**pproach Uphill and Upwind

   ☐ Monitor **R**adio Reports

   ☐ Review **P**re-incident plans

2. **Upon Arrival** – Position ICU CAR PPE - SUB

   ☐ Position vehicle **U**pwind (uphill)

   ☐ Establish Command (**IC**), or

   ☐ Identify Conditions, Actions, Resources (**CAR**)

   ☐ Ensure all members wear **PPE** and respiratory protection

   ☐ Inform in coming units of a **S**afe approach. (take the onramp, which is uphill from the accident)

   ☐ Prepare for a **U**nified command.

   ☐ Use **B**inoculars to size-up the incident.

3. **Initial Radio Report** – FM Radio

   ☐ Establish Tactical Radio **F**requency.

   ☐ Determine/Announce Operational **M**ode: (Combination: Rescue, Contain spill, Evacuate highways and possibly mall = Offensive / Confine leak, All Hot Zone Operations = consider defensive precautions until hazmat team arrives).

   ☐ IRR: Conditions, victims, size up concerns, name and location of the CP (Rt 10 Command, Upwind from accident on Rt 10 – give closest street or mile marker).

4. **Size-Up** – COAL TWAS WEALTHS

- ☐ Apparatus/staffing: will need additional resources to ensure safety on the highway (flammable liquid) and at the mall downwind (cloud).

- ☐ Life hazard, Terrain, Street conditions: high life hazard – the driver will need to be extricated and people are stuck in traffic downhill with leaking liquid. They will need to be rerouted, be aware of potential ignition sources.

- ☐ Exposures, Area: the shopping mall downwind is a concern due to the cloud. The school is not as much of a concern because of the holiday and fact that it is upwind.

- ☐ Time of day: heavy traffic on a holiday.

5. **Resources** – A RULES WAR HOG+

- ☐ Additional alarm (2nd) - to ensure an adequate amount of personnel and resources

- ☐ Rapid Intervention Crew (RIC) - for firefighter safety

- ☐ Utility Companies (gas, electric, water) – will not be needed at this time

- ☐ Law enforcement - for traffic/scene/crowd control – shut highway and divert traffic

- ☐ EMS (BLS/ambulance, ALS/paramedic) - for patient triage and treatment

- ☐ Safety Officer (SO) - for scene safety

- ☐ Water Supply Officer (WSO) – possibly, to ensure adequate water

- ☐ Accountability Officer - to track and account for all on scene personnel

- ☐ Rehab Unit – for rest and rehabilitation

- ☐ Hazardous Material Team

- ☐ Foam Tender

- ☐ Office of Emergency Management (OEM),

- ☐ Govt.: Environmental Protection Agency, etc.

- ☐ (+) Attempt to notify shipper/receiver regarding driver and contents

- ☐ Heli-EMS to ensure driver can be transported to hospital (if needed)

6. **Strategy / Tactics** – Implement 8 Step Process for Haz-Mat (SHIP IRDT)

1. **S**ite Management and Control

   ☐ Establish a security perimeter by isolating and denying entry to the area. Have law enforcement officials assist you with this responsibility.

   ☐ Establish control zones for the incident site: Hot, Warm, and Cold.

   ☐ Identify and announce a safe approach for incoming resources.

   ☐ Establish staging (Uphill and Upwind) as a method of controlling arriving resources.

   ☐ Conduct size up to determine the need for immediate rescue.

   ☐ Implement public protective actions, including: shut the highway, evacuate affected areas, and re-route traffic.

2. **I**dentify the Problem – gather date utilizing the following:

   - Placards and labels – marking and colors.

   - Shipping papers for product information, if you can access them.

   - Monitoring and detection equipment.

   - Pay attention to what you see, hear, and smell.

3. **H**azard & Risk Evaluation - gather data from the following:

   ☐ DOT Response Guidebook (reference material).

   ☐ CHEM-TREC (Technical information center).

   ☐ Hazardous material databases (online, which can be accessed via phone or laptop).

   ☐ Monitoring instruments.

   ☐ Extent/damage to visible containers.

   ☐ Predict likely behavior of the released material and containers.

   ☐ Analyze HAZARDS & RISKS to determine the safest and most effective IAP.

4. Select proper **P**PE & Protective Clothing and ensure on scene personnel are protected.

5. **I**nformation Management and Resource Coordination (ICS) - U SIL FLOP SR

    ☐ Unified Command Staff: include law enforcement, hazmat and environmental specialists.

    ☐ Safety officer: to ensure a safe working environment.

    ☐ Information officer: to communicate with the media and concerned citizens.

    ☐ Liaison officer: to communicate with other agencies and ensure smooth operations.

    ☐ Finance officer/section: cost recovery and cleanup.

    ☐ Logistics officer/section: Supply and support efforts.

    ☐ Operations officer: Hazmat group, rescue group, decon group, etc.

    ☐ Planning officer/section: Tech specialists, soil removal/cleanup.

    ☐ Staging officer: located upwind, safe distance.

    ☐ Rehabilitation officer: also located upwind, safe distance.

    ☐ Hazmat officer to focus on the hazmat identification and containment portion of the incident.

    ☐ Establish Divisions/Groups early to enhance communications and improve accountability (Ex: Rescue = Rescue Group; Spill containment = Containment Group, etc.).

    ☐ Assign division supervisors to help you manage your span of control.

6. Implement **R**esponse Objectives

    ☐ Offensive Operations: Rescue & Extrication, containment of the leak, Evacuation of personnel on highway and possibly at the mall. (Combination: Rescue, Contain spill, Evacuate the highway and possibly the mall. = Offensive / Confine leak and all Hot Zone Operations = consider Defensive precautions until hazmat team arrives).

*Identify more specific actions:

- [ ] Rescue—extricate driver, remove to safe location, air lift to hospital if necessary.

- [ ] Spill control—Dike and dam, confine and contain leaking fluids.

- [ ] Fire control— control ignition sources, especially vehicles downhill – if they cannot be removed.

- [ ] Public protective actions—divert traffic, or remove civilians from danger, monitor air quality downwind, evacuate mall if necessary (school should be closed on this holiday, but confirm that no custodians or workers are on the premises).

- [ ] Cleanup – Utilize environmental specialists.

*Assign Engine and Ladder companies:

Engine Company Ops will include:

- [ ] Position apparatus – Uphill/Upwind/Safe distance.

- [ ] Establish a primary water supply (may need to rely on shuttling water).

- [ ] Choose the appropriate size hose lines.

- [ ] Advance and position two hose lines – one to protect the driver, another to help divert leaking fluid, if necessary.

- [ ] Confine & Extinguish any fire.

- [ ] Protect life hazards.

- [ ] Assist the rescue company with extrication of the driver.

- [ ] Protect exposures.

- [ ] Utilize Thermal Imaging Cameras (TIC) to determine levels in containers.

- [ ] Coordinate with ladder companies (and other companies on scene).

- [ ] Provide periodic progress reports.

Ladder Company Ops: LOVERS-UPS (TIC) – address necessary ladder company duties

- ☐ Apparatus Placement – Uphill/Upwind.

- ☐ Primary Search of the area to ensure there are no other victims.

- ☐ Rescue Operations – assist the rescue team with extrication.

- ☐ Utilize TIC's.

- ☐ Coordinate with engine companies (and other companies on scene).

- ☐ Provide periodic radio reports.

7. Decontamination

- ☐ Assume everything coming out of the "HOT ZONE" was exposed and contaminated.

- ☐ Establish a Decontaminatin site/officer and Group.

- ☐ Conduct DECON and medical monitoring on all members exposed to hot/warm zone.

8. Terminate the incident (see #9 for steps to take)

7. **Benchmarks** – Time, Progress, RER/LIP

- ☐ Time Management (Progress reports every 10-15 min)

- ☐ RER/LIP

8. **Under Control** – A SOS PAR ICS DO

- ☐ Announce via radio

- ☐ Secondary Searches of the area

- ☐ Salvage and Overhaul responsibilities

- ☐ PAR

- ☐ Investigation Unit

- ☐ CO/Air quality levels

- ☐ Secure the Area

       ☐   Demobilize

       ☐   Turn the vehicle over to the On scene authority - police

9. **Terminating/Transferring Command** – TD DOP

       ☐   Transfer to law enforcement (traffic accident)

       ☐   Debriefing

       ☐   Document/reports

       ☐   Offer CISD

       ☐   PIA

Always Implement a Haz-Mat Plan

BONUS!

Whenever confronted with a hazardous materials release of any sort, you should implement a haz-mat response plan. Again, the 8 step plan outlined above is just one example. The important thing to remember is that it does not matter how large or small the incident is, if you follow your plan, you will address all the necessary actions needed to mitigate the incident.

Additionally, you can cut and paste your action plan into your reporting system as a foundation for your narrative when documenting the incident.

Non-Fire Scenario #5
School Bus Accident

Photo Courtesy Of: FirstDueFirePhotos.com

You are acting as one of your department's two Battalion Chiefs. On a brisk day in March, you respond to a reported vehicle accident involving a bus and arrive on scene, along with the Tour Commander (the Deputy Chief), two engine companies, and one ladder company. Upon assessing the situation, you discover the incident is far worse than you had anticipated. This is a multiple vehicle accident, involving three cars and a school bus full of ninth graders. One of the vehicles slammed into a nearby house, causing a combination vehicle and structure fire. The IC establishes command and directs the Engine and Ladder companies as they pull the vehicle's driver from the wreckage, and begin to fight the fire.

There are multiple injuries on the school bus, including the driver and several student passengers. The IC designates you as the Rescue Operations officer and directs you to take charge of and mitigate the school bus portion of the accident.

Scenario Question

As the officer in charge of the "school bus" portion of this incident, what actions would you take?

Non-Fire Scenario #5 - Answer Key

1. **Acknowledge your assignment (always via radio and/or in person)**

2. **Size-Up the scene**

   ☐ Assess the scene and surrounding area for safety hazards.

   ☐ Determine the number of victims, and account for all passengers.

   ☐ Observe street conditions, and secure the area from onlookers and traffic.

   ☐ There will be a large fuel tank. The driver may help you determine how much fuel is in it.

     – Determine if there is a fuel spill and if it can be contained.

   ☐ Assess the possibility of a fire hazard, and the options for water supply.

   ☐ Determine if the time of day or weather conditions affect operations.

     – Are the roads wet and slippery in a heavy traffic area?

     – Do the victims need to be relocated to a warmer area?

   ☐ Determine if there are any threatened exposures.

3. **Provide an Initial Radio Report**

   ☐ Report observations, provide information about injured passengers, and call for resources.

4. **Resources needed for your portion of the incident include:**

   ☐ Additional alarm, to ensure an adequate number of personnel (confirm w/ IC).

   ☐ Law enforcement for scene safety, traffic, crowd and media control.

   ☐ Adequate number of ambulances/EMT's for treatment, triage, and transportation of victims.

   ☐ A representative of the school for victim tracking.

☐ A heated bus (or heated location) for victims during cold weather incidents.

☐ A Public Information Officer (to provide info to the media and parents) will need to be assigned early. Make this suggestion to the IC.

5. Develop an Incident Action Plan (IAP)

☐ Ensure that all on scene personnel are aware of the goals and objectives of the IAP.

6. Conduct Rescue Operations

☐ Establish a water supply.

☐ Secure the area by establishing a safe zone (use caution tape).

☐ Activate your department's mass-casualty incident (MCI) plan.

- This will require additional agencies, supplies, and resources to be called to the scene.

☐ Notify the local hospital or medical facility of the number of patients being transported.

☐ If it's dark, set up lights and use a TIC to search for victims who may have been extracted from the bus or other involved vehicle(s).

☐ Ensure that all personnel working in the danger zone are wearing proper PPE.

☐ Stretch a hose line to the bus so it is in position in the event of fire.

☐ Disconnect batteries from the electrical system; have the bus driver assist with directions.

☐ Stabilize the vehicle.

☐ Remove passengers through the windows if the doors are not accessible.

☐ Set up a medical evaluation area outside the Hot Zone.

- Staff this area with EMS personnel, assign an EMS officer, and set up a triage system.

☐ Assist walking wounded from the danger zone to the medical evaluation area.

☐ Extricate victims who are trapped, and remove them to the medical evaluation area.

☐ Assign a victim tracking coordinator. This may be an uninjured bus driver or a school representative.

☐ Provide essential information to the PIO.

☐ Have a decontamination area set up if victims have been contaminated with fuel.

☐ Rotate personnel frequently if the incident is long in duration.

7. **Upon completion, take the following actions:**

☐ Clean up the scene. (if needed, call for a cleanup contractor).

- For small spills, use absorbent or sand.

- For large spills and when ice is on the roadway, call a municipal agency for use of a vehicle like a salt or sand spreader.

- Notify the Department of Environmental Protection or the Environmental Protection Agency of the spills, and include their case number in your report.

- Sweep up glass, and remove any large (or sharp) debris from the roadway.

- If this is a suspected crime scene, preserve evidence and call for a fire investigator.

- Do not leave until the vehicle is removed from the scene and roadway is safe.

☐ Demobilize the incident.

☐ Ensure the scene is safe.

☐ Report to the IC and brief him/her on the actions you have taken.

☐ Suggest critical incident stress debriefing for personnel.

Note: All states require that you report crashes involving a truck or bus, with defined severity criteria, to the Motor Carrier Management Information System (MCMIS) crash file, maintained by the Federal Motor Carrier Safety Administration FMCSA Support Services.

Prepare for Your Most Likely Scenarios

School and public transportation bus accidents offer a wide range of challenges, most of which revolve around the removal and treatment of occupants. At bus accidents, all the tactics normally used at car accidents apply; however, extra emphasis should be placed on victim tracking and scene safety, especially if the accident involves an occupied school bus. Having a school representative respond and assigning a public information officer (PIO) is essential, considering the fact that you will have concerned and frantic parents calling in or trying to come to the scene to account for their children.

Although school bus incidents would be considered low-frequency, they are also high-risk. Additionally, school buses are the biggest type of mass transit in the United States, providing almost nine million student trips every year. In comparison, this is twice as many passenger trips than provided by transit buses across the nation. According to the U.S. Department of Transportation (USDOT) and the National Highway Traffic Safety Administration (NHTSA), there are more than 50,000 school bus accidents every year in the United States. This is a scenario worth preparing for.

On that note, think about incidents that are happening in your area and prepare for them. In Detroit, Michigan and Camden, NJ, for example, they are seeing an increasingly high amount of vacant building fires. If you lived in or near these areas, it would only make sense to practice scenarios involving vacant buildings.

Non-Fire Scenario #6
Incident Rehabilitation

Photo By: Andrew Taylor

The members of your department, and several other mutual aid companies, have been battling a four alarm fire in a large warehouse on Sullivan Street for several hours. You are one of two battalion chiefs on scene. You have just been called to the scene because you were finishing up at another job across town. Shortly after your arrival, the Incident Commander informs you that you, and one Engine company crew, will be establishing a rehabilitation group. He gives you the assignment, and tells you to "get it done asap," because several members have been showing signs of heat exhaustion.

Scenario Question

As the Battalion Chief, what actions would you take to set up this group?

(Include information on how you would provide treatment for firefighters who enter the rehabilitation area)

1. **Announce/confirm your assignment and location over the radio.**

   ❑ You have been assigned as the rehab officer/supervisor.

2. **Designate a site/location – taking the following into consideration.**

   ❑ Ensure one entry/exit point, for accountability.

   ❑ Protection from weather elements.

   ❑ Away from exhaust fumes of apparatus and equipment.

   ❑ Far enough away for firefighters to remove their gear, relax, and get a mental and physical rest from the stressful incident.

   ❑ Easily accessible by EMS vehicles and personnel.

   ❑ Large enough area to accommodate multiple crews.

   ❑ Allow for easy access to and from the incident scene.

   ❑ Possible choices for site location include:

   > ➤ A nearby structure, garage, or building lobby.

   > ➤ An emergency vehicle (fire apparatus, ambulance, other) with heat or AC.

   > ➤ A tent (or tarps) set up in an open area.

   > ➤ A school bus or public transportation vehicle.

   ❑ Use fire or caution tape to identify the entrance and exit of the rehabilitation area.

3. **Call for resources.**

   ❑ Staff adequately, request additional qualified EMS personnel if needed (1:5 ratio).

   ❑ EMS personnel should be trained at the minimum level of basic life support, preferably the EMT level or above.

   ❑ A supply of fluids (water, activity beverage, oral electrolyte solutions, and ice).

   ❑ A supply of food (snacks, soup, broth or stew in hot/cold cups).

- Medical supplies (blood pressure cuffs, stethoscopes, oxygen administration devices, cardiac monitors, intravenous solutions and thermometers).

- Other supplies (awnings, fans, tarps, smoke ejectors, heaters, dry clothing, extra equipment, floodlights, blankets and towels, traffic cones, and fire-line tape).

4. **Rotate companies/personnel frequently.**

- Ensure that fire companies rotate frequently and enter and exit together, maintaining crew integrity.

- Crews should remain intact and tag in and out to assure accountability.

- Ensure adequate and equal rest for all personnel operating at the scene (minimum rest time should be 15 minutes).

- Rehabilitation time should be increased for individuals who need additional rest and monitoring.

- The two-air-bottle rule, or 45 minutes of work-time, is recommended as an acceptable standard prior to mandatory rehabilitation.

5. **Triage and treatment of firefighters would be based on the following guidelines:**

- Evaluation and treatment would include.

  - Take and monitor vital signs.

  - Heart rate—measure for 30 seconds; if heart rate exceeds 110 beats per minute, an oral temperature should be taken.

  - Temperatures—if the member's temperature exceeds 100.6F, he/she should not be permitted to wear protective equipment; if it is below 100.6F and the heart rate remains above 110 beats per minute, rehabilitation time should be increased; if the heart rate is less than 110 beats per minute, the chance of heat stress is negligible.

  - Hydration—Provide a rehydration solution that's a 50/50 mixture of water and a commercially prepared activity beverage. Avoid alcohol, caffeine, and carbonated beverages.

  - Nourishment—Provide recommended foods such as a cup of soup, broth, or stew, protein or energy bars, apples, oranges, or bananas; avoid fatty and/or salty foods.

  - Allow for mental and physical rest (minimum rehab time should be 15 minutes).

> Medical treatment should be provided in accordance with acceptable control procedures.

❑ Documentation and monitoring

> Use a standard medical evaluation form for each firefighter (include name, date, time in, members complaint, vitals, medical treatment provided, and time out).

> Monitor firefighters to see whether their conditions improve, maintain the same or deteriorate and treat accordingly.

> Medical evaluations are confidential and should be given to the individual upon termination of treatment.

> Monitor for signs or symptoms of dehydration.

> Monitor for signs or symptoms frostbite (if cold).

> Monitor for signs of heat stroke (if hot).

> Monitor for signs or symptoms of other potential problems.

6. **Track personnel to ensure accountability.**

❑ Crew member names should be documented in a rehab entry form.

❑ Crew entry and exit times should be documented.

❑ Tag in and out using personal accountability tags (PAT).

❑ Evaluation and treatment should be documented.

❑ Crews should not exit the rehab area unless released by you or a designated the rehab supervisor.

7. **Utilize medical charts/forms at a rehab site as guidelines, such as:**

❑ Temperature (heat/cold weather) guidelines

❑ Rehab entry forms

❑ Individual medical evaluation forms

Preparation Before Presentation

I was working as the Incident Commander of a two-alarm residential structure fire in my community when a Battalion Chief from a neighboring department arrived on the scene with one truck and two engine companies. He approached me at the command post to check in and get his assignments, but before I said a word he enthusiastically said, "I read the book!"

My focus was on the fire, as it should have been, so I gave him the assignments for his companies to carry out and we proceeded to put out the fire. A while later, the BC approached me again and repeated the comment he made earlier.

"I read the book." he said, before adding, "I am a BC today because of your book."

"Which book?" I asked.

"Practice Scenarios." he replied.

During our brief conversation, he informed me that he felt he had the experience and knowledge, he just didn't know how to package it nicely in the assessment center. That's what 'Practice Scenarios' gave him. Once he had a system, he had the missing ingredient. The only thing left to do was prepare.

When it comes to taking promotional exams, preparing to present your answer is vitally important. You don't want to make the mistake that so many others do, which is showing up on examination day unprepared. If the first time you give an oral presentation is on the day of your test, you are not playing your hand well. What exactly are you preparing? In the shortest possible answer, you are preparing to hit the main points that will provide you with the best possible score, and you are preparing to deliver your message in the most professional manner possible. This entire book was constructed to help you understand the preparation process.

On another note, after the same fire I referenced above, someone posted several on-scene photos of us on social media. Under a photo of me, one person commented. "This was the first guy to show up. He was moving quickly and certain of purpose." That comment sums up another benefit that the information in this book provided me with - confidence. Never underestimate the value of systems and preparation.

Non-Fire Scenario #7
Compounded Incident

Photo By: Ron Jeffers

It's August 1st at approximately 1030 hours when you respond to a reported structure fire at Hamilton Warehouse, an old, three-story, ordinary warehouse that has been known to store pool chemicals in the past. The company name and ownership has changed so often over the past few years that neither you, nor your crew, know what products are currently stored within the warehouse. The building is approximately 300' by 120'. Three sides of the building present no exposure threat. On the B side is an attached three story office complex with 8 business names on the door.

Two engine companies, two ladder companies, and a BC are dispatched and arrive on the scene before you. The first Engine arrives on the scene and reports a heavy smoke condition inside the building. They stretch a 2 ½-inch line into the structure. Several workers exited the building saying there was an explosion in the C/D corner of the building, which according to one of the employees, is the section they now use to store flammable liquids.

As you arrive on scene, a second line is being stretched into the building from Engine 2. Ladder 1 reports that fire is showing from the C/D corner of the building. Ladder 2 crew is inside performing an interior search of the first floor. As you exit your vehicle, a second explosion occurs, causing partial collapse of the C/D side of the building. This is confirmed by Ladder 1personnel who are making their way off the roof and onto the ladders to retreat to safety. You do not receive a mayday, but you know Engine 1 and Ladder 2 personnel are working in that area. It is 98 degrees with a slight wind blowing towards the East (D) side of the building.

Scenario Question

Describehe actions you would take from the initial call until termination of this incident.

1. **En route** – WARP

   ☐ **W**ind speed/direction and projected weather

   ☐ **A**pproach Upwind

   ☐ Monitor **R**adio Reports

   ☐ Review **P**re-incident plans

2. **Upon Arrival** – Position ICU CAR PPE - SUB

   ☐ Position vehicle **U**pwind (uphill)

   ☐ Face-to-face with current **IC**

   ☐ Incident Debriefing - Conditions, Actions, Resources (**CAR**)

   ☐ **A**ssume Command

   ☐ **R**e-assign IC to operations or another position (hazmat officer)

   ☐ Ensure all members wear **PPE** and respiratory protection

   ☐ Inform in coming units of a **S**afe approach (if needed)

   ☐ Prepare for a **U**nified command (with hazmat and rescue specialists)

   ☐ Use **B**inoculars to size-up the incident (if needed)

3. **Initial Radio Report** – FM Radio

   ☐ Establish two Tactical Radio **F**requency's (the initial frequency is for the FF rescue – if trapped in collapse. The second frequency is for fire/hazmat operations).

   ☐ Determine/Announce Operational **M**ode: Combination – Offensive for rescue, Defensive in the collapsed portion of the bldg. (if rescue ops are not needed).

   ☐ IRR: Conditions, victims, size up concerns, name and location of the CP (Hamilton warehouse Command, A side of the building).

   ☐ Call for a PAR to determine if members are trapped.

4. **Size-Up** – COAL TWAS WEALTHS

   ☐ Construction: Large warehouse, partial collapse, implement a collapse response plan.

   ☐ Occupancy: warehouse with flammable liquids and other chemicals, implement a haz-mat plan.

   ☐ Apparatus/staffing not sufficient, call for additional alarms, rescue, hazmat team, etc.

   ☐ Life hazard: high (FF's, Employees of the warehouse and Employees of attached exposure must be accounted for. Consider the time of day).

   ☐ Auxiliary appliances: Sprinkler system can be supplied, but review MSDS sheets (for chemical reactivity to water) and ensure FF's are not trapped before charging lines.

   ☐ Weather: hot - Rehab and rotation of members will be essential.

5. **Resources** – A RULES WAR HOG+

   ☐ <u>A</u>dditional alarms – 3rd/4th - to ensure adequate personnel and resources.

   ☐ <u>R</u>apid Intervention Crew (RIC) - for firefighter safety .

   ☐ <u>U</u>tility Companies (gas, electric, water) – for utility control.

   ☐ <u>L</u>aw enforcement - for traffic/scene/crowd control.

   ☐ <u>E</u>MS (BLS/ambulance, ALS/paramedic) - for patient triage and treatment.

   ☐ <u>S</u>afety Officer (SO) - for scene safety (Assign 2 Safety Officers).

   ☐ <u>W</u>ater Supply Officer (WSO) – to ensure adequate water.

   ☐ <u>A</u>ccountability Officer - to track and account for all on scene personnel.

   ☐ <u>R</u>ehab Unit – for rest and rehabilitation.

   ☐ <u>H</u>azardous Material Team.

   ☐ <u>O</u>ffice of Emergency Management (OEM), to help acquire hazmat resources and heavy machinery for moving debris.

   ☐ <u>G</u>overnment agencies - Environmental Protection Agency.

   ☐ (+) USAR, Rescue Co.

   ☐ Structural Engineer.

- ☐ Air supply/mask service unit.
- ☐ Lighting for extended operations.

6. **Strategy / Tactics** – Combine firefighting tactics + 8 Step Hazmat Plan (SHIP IRDT) + structural collapse response plan

*Firefighting tactics

Engine Company Ops:

- ☐ Position apparatus upwind, out of collapse zone, but in area where you can initiate attack.
- ☐ Establish a primary (and secondary) water supply .
- ☐ Initiate attack if deemed possible (after assessing hazmat & collapse situations).
- ☐ Choose appropriate size hose lines.
- ☐ Advance and position hose lines to the fire and to the collapse area to assist with rescue operations (if possible).
- ☐ Locate, Confine, & Extinguish (LCE) the fire if possible. If not, prepare for large fire and/or defensive operation.
- ☐ Protect life hazards.
- ☐ Protect exposures using hose lines.
- ☐ Supply auxiliary appliances, but don't charge the systems until you are sure the haz mat products will not have an adverse reaction to water.
- ☐ Utilize Thermal Imaging Cameras (TIC).
- ☐ Coordinate with ladder companies (and other companies on scene).
- ☐ Provide periodic progress reports.
- ☐ Eliminate all ignition sources as you go.

Ladder Company Ops: LOVERS-UPS (TIC-COP)

- ☐ Position apparatus to ensure all personnel on the roof have means of egress -

Raise the aerial to allow for an elevated view to survey the site, and set up for elevated master stream operations.

- ☐ Raise and position <u>L</u>adders (if needed) for egress and firefighting ops.

- ☐ Force <u>E</u>ntry where needed.

- ☐ Conduct a primary <u>S</u>earch for missing or trapped firefighters (and later, a secondary search).

- ☐ <u>R</u>escue missing or trapped firefighters – remove any civilians from danger (including attached office building).

- ☐ <u>V</u>entilate if needed in an area where firefighting/rescue operations are being conducted.

- ☐ <u>U</u>tility control should be from the outside, via the utility companies.

- ☐ Salvage and <u>O</u>verhaul should be conducted if necessary.

- ☐ Utilize <u>TIC</u>'s to assist with operations.

- ☐ <u>C</u>oordinate with engine companies (and other companies on scene).

- ☐ Provide periodic <u>P</u>rogress reports.

- ☐ + Eliminate all ignition sources as you go.

*<u>Implement your Collapse Rescue Plan</u>

- ☐ Conduct a personnel accountability Roll Call to determine how many firefighters (if any) may be hurt, trapped or unaccounted for.

- ☐ Conduct a quick collapse size-up, consisting of the following.

  - ☐ Construction: Identify any construction features that could promote further/secondary collapse.

  - ☐ Street: Be aware of any truck or train traffic in the immediate area.

  - ☐ Weather: Call for projected weather report because it may have an impact on the operation and overall safety of the incident.

  - ☐ Exposures: Identify any surrounding or attached properties that might be affected by the collapse.

  - ☐ Area: Conduct a 360-degree view of the incident (if possible) to determine if the collapse is extensive or localized.

  - ☐ Life Hazard: Determine if any civilians are unaccounted for.

131

- ➢ Perform a risk analysis: are you confronted with a rescue or a recovery?

- ☐ Exposures: Determine the collapse type:
  - ➢ Lean-to collapse
  - ➢ Pancake collapse
  - ➢ Unsupported collapse
  - ➢ V-type collapse
  - ➢ Inward/outward collapse
  - ➢ 90-degree collapse

*Preparing for collapse rescue operations would include the following:

- ☐ <u>Site management and control</u> (all dependent upon the hazmat situation at hand):

  - ☐ Establish a security perimeter by isolating and denying entry to the area/building. Have the police to assist.

  - ☐ Remove all non-essential personnel and control the scene.

  - ☐ Evacuate all affected areas and any attached or nearby structures.

  - ☐ Conduct a site survey. Gather information specific to risk analysis, the location of trapped individuals, what might have caused the collapse, secondary collapse concerns, and accessibility.

  - ☐ Compile all information that will assist with the rescue operation and safety of the incident.

  - ☐ Establish collapse and operational zones (at least 1.5 times the height of the building).

  - ☐ Utility control: eliminate an explosive atmosphere, electrocution, and a potential drowning.

  - ☐ Stretch protective hose lines.

  - ☐ Shore unstable areas.

  - ☐ Conduct air monitoring.

  - ☐ Use surveyors transit to detect early structural movement/secondary collapse.

☐ Eliminate vibrations.

☐ Increase supervision in difficult and dangerous areas.

☐ Increase lighting in area.

☐ Conduct frequent PARS.

☐ Conduct frequent reliefs.

☐ Provide food and shelter.

☐ <u>Conduct the actual rescue operations</u>

    ☐ 1. <u>Surface victim</u> search and removal.

    ☐ 2. <u>Void search rescue</u>.

    ☐ 3. <u>Selected debris removal</u> and tunneling.

    ☐ 4. <u>General debris removal</u> and search - This phase of the operation would require additional resources, such as: Outside agencies with heavy construction equipment/operators, specially trained cadaver dogs, on scene/ critical incident stress debriefing (CISD).

*<u>Implement 8 Step Plan for tactical management of hazmat incidents (SHIP IRDT)</u>

1. <u>**S**ite Management and Control</u>

   ☐ Establish a security perimeter by isolating and denying entry to the affected area (except for essential rescue personnel). Give significant consideration in using the police in assisting you with this responsibility.

   ☐ Establish control zones for the incident site: Hot, Warm, and Cold.

   ☐ Assure a safe approach for incoming resources.

   ☐ Establish staging (Upwind) as a method of controlling arriving resources

   ☐ Implement public protective actions by evacuating the warehouse and attached building, protecting in place, or implementing a combination of the two.

2. <u>**I**dentify the Problem</u>

   ☐ Determine building occupancy and location; the name on the front of the building may tip you off to specific concerns.

133

- ☐ Container shape can indicate pressurized vs. non-pressurized containers.

- ☐ Marking and colors as well as placards and labels can provide product information and mitigation options to first responders.

- ☐ Facility documents and Material Safety Data Sheets (MSDS) will provide product information.

- ☐ Monitoring and detection equipment can identify the presence of a product.

- ☐ Your senses can help identify the problem. Pay attention to what you see, hear, and smell.

3. <u>Hazard & Risk Evaluation</u>

Gather hazard data from the following:

- ☐ Reference materials (Department of Transportation Emergency Response Guidebook – DOT/ERG).

- ☐ Technical information centers such as CHEM-TREC.

- ☐ Hazardous material databases.

- ☐ Right-to-know information; Material Safety Data Sheets (MSDS).

- ☐ Monitoring instrument.

- ☐ Determine the extent of damage to container.

- ☐ Predict the likely behavior of the released material and containers.

- ☐ Analyze HAZARDS & RISKS to determine the safest and most effective Incident Action Plan

4. <u>Select proper PPE & Protective Clothing</u>

- ☐ Is HazMat gear/equipment needed, or will firefighting gear/respiratory protection be enough?

5. <u>Information Management and Resource Coordination (ICS)</u> - U SIL FLOP SR

- ☐ Unified Command Staff (Fire, Police, HazMat, Rescue, EMS).

- ☐ Safety officers (to ensure a safe working environment).

- ☐ Information officer (to communicate with the media and concerned citizens).

- ☐ Liaison officer (to communicate with other agencies and ensure smooth operations).

- ☐ Finance officer/section.

- ☐ Logistics officer/section.

- ☐ Operations officer.

- ☐ Planning officer/section.

- ☐ Staging officer.

- ☐ Rehabilitation officer.

- ☐ Assign victim tracking coordinators to account for employees of the fire building and the attached exposure.

- ☐ Additional: Establish Divisions/Groups early (example: Rescue = Rescue Group).

- ☐ Assign division supervisors.

6. Implement **R**esponse Objectives

- ☐ Rescue – top priority.

- ☐ Defensive operations—confinement – after all personnel are out and accounted for.

- ☐ Spill and Leak control – only if possible to do from safe location.

- ☐ Fire control—(See Engine and Ladder company operations).

- ☐ Implement public protective actions—for example, evacuate the occupants of the warehouse and attached building.

7. **D**econtamination

- ☐ Assume anything coming out of the "HOT ZONE" is exposed and potentially contaminated.

- ☐ Establish a decon site/officer and Group.

- ☐ Conduct DECON and medical monitoring for all members.

8. Terminate the incident (see #9 for steps to take).

7. **Benchmarks** – Time, Progress, RER/LIP

   ☐ Time Management (Progress reports every 10-15 min)

   ☐ RER/LIP

8. **Under Control** – A SOS PAR ICS DO

   ☐ Announce via radio

   ☐ Secondary Searches

   ☐ Salvage and Overhaul responsibilities

   ☐ PAR

   ☐ Investigation Unit

   ☐ CO/Air quality levels

   ☐ Secure the Building/Area

   ☐ Demobilize

   ☐ Turn the building over to the Owner or on scene authority

9. **Terminating/Transferring Command** – TD DOP

   ☐ Transfer or Terminate

   ☐ Debriefing

   ☐ Document/reports

   ☐ Offer CISD

   ☐ PIA

Compounded Incidents & Redundancy

Compounded incidents can be very intimidating, but the remedy for intimidation is preparation. These incidents will call for the merging of response plans. Some activities you will do for collapse will not need to be repeated when explaining what actions you will take for fire suppression and hazmat response; however, redundancy trumps an inadequate response.

Supervision Scenarios

The Subordinate Interview

High ranking officers in management positions must learn to deal with subordinate issues like conflict resolution or sub-standard performance. There is no shortage of information teaching you how to conduct a formal subordinate interview and address serious problems, but this section shares the essential steps of conducting such an interview.

Here are the basic components of a subordinate interview as it would occur within the fire service in a simple, numbered list.

1. Gather facts – Review all pertinent info (Including the personal files of all individuals who are involved in, or who witnessed the incident at hand. Also review incident reports).

2. Get it in writing – Have all involved provide written reports of what occurred and why.

3. Schedule a meeting – If you suspect that disciplinary action will have to be taken, the individual(s) should be provided with the option of having union or legal representation.

4. Conduct the Meeting – Begin by putting the firefighter at ease. State the purpose of the meeting. Discuss positives about the individual before stating the problem at hand. When you do discuss the problem, use facts to support the reason why this was brought to your attention and explain why it's inappropriate behavior.

5. Get the individual's side of the story – Use open-ended questions to gather facts. Probe for answers, and show empathy.

6. Look for an underlying problem. Don't discard the possibility that circumstances in the firefighter's personal life may be affecting his/her behavior, especially if it is uncharacteristic.

7. Determine which of the 3U's you are dealing with (Unaware, Unable or Unwilling) - The actions you take will be determined by which of the 3U's the individual falls under.

8. Discuss progressive discipline (verbal reprimand, written reprimand, suspension, fines, termination) – Any time you take disciplinary action, you should explain what the next step may be if another issue arises.

9. Develop a solution together and implement it – You will achieve better results if FF's feels like they are playing an active role in determining what actions can be taken to correct the situation. When discussing solutions, take training & counseling into consideration: Training (NFPA, organizational procedures, sensitivity training, specific training); Counseling (Employee Assistance Programs, Critical Incident Stress Debriefing, etc.)

10. Summarize – recap the key points and the solution to ensure clarity.

11. Set a follow up meeting – stress the fact that improvements need to be made within the specified time frame. Whenever appropriate, a strong leader will also ensure that what occurred in the meeting will remain confidential and that you have an open-door-policy if the individual needs to discuss related issues.

12. Inform the individual of the appeal process – Discuss 'Due process' if they feel the actions taken to correct the situation were too harsh.

13. Close on a positive note.

14. Document and report – document in writing what occurred in the meeting and inform your superior of such. Also inform complainants that corrective actions have been taken.

15. Monitor and evaluate the individual's progress until the next meeting.

The 3 U's

The goal of a fact-finding interview is to determine what the problem is and to attempt to develop a solution. This can only be done after concluding if the individual is **Unaware**, **Unable** or **Unwilling** (otherwise known as the 3U's). If you fail to make this determination – whether informally, or formally – you will not be able to take the appropriate corrective actions.

Here is a brief description of the 3U's:

✓ Unaware: Not aware or not conscious of what is going on.

✓ Unable: Lacking mental or physical capability or efficiency; incompetent.

✓ Unwilling: Boldly resisting authority or having a defiant attitude. (Insubordinate).

Consider it your job, as an officer, to determine if the individual is unaware that there is a problem, unable to fix it, or unwilling to fix it. As you gather facts, you should begin contemplating your course of action. You can do this by thinking:

If he/she's *Unaware*, I will…

If he/she's *Unable*, I will…

If he/she's *Unwilling*, I will…

After determining which category the subordinate falls into, implement the appropriate solution, which may include some form of training, counseling, and/or a variety of other possibilities – all depending on the issue(s) at hand.

Supervision Scenario #1
Uncharacteristic Behavior

A third year firefighter has been displaying uncharacteristic behavior from what he has become known for. He is considered to be an ambitious and enthusiastic firefighter, however, over the past few weeks he has been showing up late for work, performing below standard, and complaining frequently about various issues – in the firehouse and on the fire ground. He has also been spending a lot of time on his cell phone, and can often be heard arguing with the person on the other end of the line. After receiving complaints from other firefighters, you meet with the firefighter to address the situation.

Scenario Questions

1. What information will you gather before the meeting?

2. How will you conduct this interview?

1. What information would you gather?

☐ Gather facts – Review all pertinent info.

  ➤ Personnel files of all individuals who are involved or who witnessed the incidents.

  ➤ Personnel file of the firefighter in question.

  ➤ Any incident reports where incidents concerning this individual were mentioned/ documented.

☐ Get it in writing

  ➤ Have the FF's who are complaining provide written reports of what has been occurring.

2. How will you conduct this interview?

☐ Schedule a meeting – inform the firefighter that he has the right to bring union representation.

☐ Conduct the Meeting.

  ➤ Begin by putting the firefighter at ease.

  ➤ State the purpose of the meeting.

  ➤ Discuss positives about the individual before stating the problem at hand.

  ➤ Use facts to support the reason why this was brought to your attention and explain why it's inappropriate behavior.

☐ Get the individual's side of the story.

  ➤ Use open-ended questions to gather facts.

  ➤ Probe for answers.

  ➤ Show empathy.

☐ Look for an underlying problem.

- Circumstances in the individual's personal life may be affecting his behavior, especially since this behavior is uncharacteristic.

☐ Determine which of the 3U's you are dealing with (Unaware, Unable or Unwilling).

- The actions you take will be determined by which of the 3U's the individual falls under (see bonus section of this answer key).

☐ Discuss progressive discipline - Verbal reprimand, written reprimand, suspension, fines, and termination.

- Explain what the next step may be if another issue arises.

☐ Develop a solution together and implement it.

- Take training and counseling into consideration:

  - Training (NFPA standards, SOP's, organizational procedures, sensitivity training).

  - Counseling (Counseling, Employee Assistance Programs, Critical Incident Stress Debriefing).

☐ Summarize – recap the key points to ensure clarity.

☐ Set a follow up meeting.

- Stress that improvements need to be made within the specified time frame.

- Open-door-policy.

☐ Inform the individual of the appeal process – Discuss due process if they feel the actions taken to correct the situation is too harsh.

☐ Close on positive note.

☐ Document and report.

☐ Monitor and evaluate the individual's progress until the next meeting.

- Document in writing what occurred in the meeting and inform your superior of such.

- Inform the firefighters who brought this to your attention that the issue has been addressed and corrective actions have been taken (do not give specifics).

☐ Monitor and evaluate the individual's progress until the next meeting.

The 3 U's - Be Specific

BONUS!

Depending on the response(s) you receive during the interview, you should be able to determine which of the 3U's you are dealing with. From there, you will want to address the root of the problem. Here is an example of how you might treat each of the three in this scenario.

If he is **Unaware** of his actions, perhaps the talk will be enough to correct the situation. You should still schedule some form of training, such as reviewing any related policies and procedures. You should also discuss what you expect from the individual and have him acknowledge, so there is no misunderstanding. The key word here is awareness. If the problem persists, you will no longer be dealing with a person who is Unaware. You will now have to address the issue based on the understanding that the individual is either Unable or Unwilling to correct the situation. After meeting of this nature, be sure to document the incident for your records.

If he is **Unable** to change his actions, you have a more serious situation on your hands. This individual will be showing signs of incompetence. It is tough to generalize how to deal with this situation, but there are some basic steps you should begin taking. First, the actions outlined in the previous paragraph must be repeated to ensure the person is, in fact, aware of what is expected of him. If it becomes clear that he is unable to fix the problem(s), or if this is a repeat offense, you will have to make your superior(s) aware of the situation. Again, your organization's policies and procedures should be followed at this point, but be sure to document the actions you have taken so far. You may have to refer to your notes at a later date. Be sensitive to the possibility that the person may be dealing with a personal issue that is leading to his inability to function at an acceptable level. If this is the case, private counseling may be the answer. Either way, if this is a repeat offense, you should have the individual write a special report explaining what is happening.

If he is **Unwilling** to change his actions, more drastic measures must be taken. In this scenario, it is assumed that you have already ruled out the possibility that the individual is *Unaware* or *Unable*. You are now dealing with someone who is boldly resisting authority, or having a defiant attitude. This person is insubordinate, and the situation must be immediately bumped up the chain of command and dealt with at a higher level. There may not be much you can do to a person who complains, but there are definite actions a department will have to take when a person consistently shows up late for work and performs below an acceptable standard. Once again, your organization's policies and procedures should be followed, and your actions should be documented. Your report, along with the special report completed by the individual, should be immediately sent to your superior.

Supervision Scenario #2
Sexual Harassment

Your department recently hired its third female firefighter. You return from a minor incident to find her sitting alone on the apparatus floor. She is visibly upset. She asks to speak with you alone in your office. Once alone, she proceeds to tell you that the senior firefighter she works with is sexually harassing her. She also tells you her boyfriend is aware of the situation and is on his way to the station to "take care of it."

Scenario Questions

1. What information do you need to sather in order to address this problem?

2. How will you handle this situation?

1. What information would you need to address this problem?

☐ Gather facts – Review all pertinent info:

  ➤ Personnel files of all individuals who are involved (Senior Firefighter and female fire-fighter).

  ➤ Interview others (officer and crew) to see if they witnessed anything.

  ➤ Personnel file of the firefighter in question.

  ➤ Any incident reports where incidents concerning this individual were mentioned/documented.

  ➤ Review your sexual harassment policy.

  ➤ Speak with the law department that represents your organization.

  ➤ Seek advice from your superior officer (Chief).

☐ Get it in writing.

  ➤ Have all firefighters involved provide written reports regarding this issue.

2. How will you handle this situation?

☐ While you have the female firefighter in your office, you can begin the interviewing process. – inform her that she has the right to bring union representation.

☐ Conduct the Meeting.

  ➤ Begin by putting the firefighter at ease.

☐ Get her side of the story:

  ➤ Use open-ended questions to gather facts.

  ➤ Probe for answers, and

  ➤ Show empathy.

- ☐ Ask what she would like to see happen.

- ☐ Immediately separate the female firefighter and the senor firefighter so they are not working together.

  - ➢ Move the accused firefighter.

- ☐ See if the female firefighter can get in touch with her boyfriend to advise him not to come to the firehouse because the situation is being taken care of. (Note: you can meet with the boyfriend if he persists, but make sure the senior firefighter is not in the firehouse when he arrives, and have a witness in your office).

- ☐ Tell her you will investigate, take actions to resolve the issue, and get back to her.

  - ➢ Open-door-policy.

- ☐ Document and report.

  - ➢ Document in writing what occurred in the meeting and inform your superior of such.

- ☐ Schedule a meeting to conduct an interview with the senior firefighter – inform Firefighter that he has the right to bring union representation.

- ☐ Conduct the Meeting.

  - ➢ Begin by putting the firefighter at ease.

  - ➢ State the purpose of the meeting – sexual harassment.

  - ➢ Discuss positives about the individual before stating the problem at hand.

  - ➢ Use facts to support the reason why this was brought to your attention and explain why it's inappropriate behavior.

- ☐ Get the individual's side of the story.

  - ➢ Use open-ended questions to gather facts.

  - ➢ Probe for answers.

  - ➢ Show empathy.

  - ➢ Get it in writing.

- ☐ If true, suspend the firefighter immediately.

  - ➢ Allow for due process, union representation.

- ☐ If unsure, look for an underlying problem.

  - ➢ Circumstances in the individual's personal life may be affecting his/her behavior, especially since this behavior is uncharacteristic.

☐ Try to determine which of the 3U's you are dealing with (Unaware, Unable or Unwilling).

➤ Regardless of which one you are dealing with, if the firefighter is guilty, he will face suspension. This, however, is important information for your report.

☐ Discuss progressive discipline - Verbal reprimand, written reprimand, suspension, fines, and termination.

➤ Explain that sexual harassment is a serious accusation, refer to policy.

☐ Offer counseling.

☐ This issue will require attention from your superiors and your organizations legal representatives and will be immediately bumped up the chain of command.

☐ Document the meeting.

☐ Get back to the female firefighter, and work on a solution that she agrees with

➤ Offer Counseling (Counseling, Employee Assistance Programs, human resources, etc.)

☐ Summarize the actions being taken – recap the key points to ensure clarity.

☐ Set a follow up meeting.

➤ Open-door-policy

➤ Stress confidentiality

☐ Close on a positive note.

☐ Document and report.

➤ Document in writing what occurred in the meeting and inform your superior of such.

➤ Inform the firefighters who brought this to your attention that the issue has been addressed and corrective actions have been taken (do not give specifics).

☐ Monitor and evaluate the female firefighter's progress until the next meeting.

☐ Speak with the head of your organization and suggest/request sexual harassment training for all department members.

☐ Establish a policy on sexual harassment stating department disapproval and intent to take appropriate disciplinary action against violators.

Sexual Harassment Policy

According to the Fire Chief's Handbook, Sexual Harassment Policies should cover the following information.

1. Definition (of what type of acts constitute sexual harassment).

2. Complaint (procedure explaining how and whom the individual should report alleged harassment).

3. Supervisors Responsibility (actions needed, and details on whom they report the complaint too).

4. Investigation (mention who will conduct the investigation).

5. Disciplinary Action (list what types of actions may be taken based on the severity of the harassment).

6. Prevention (explain management's responsibility to notify workers of the policy and enforce procedures).

7. Communicate policy to all employees (post it on walls at fire stations).

8. Document and record all charges.

Your previous successor was a classic micro-manager. As a new officer, you do not want to make the same mistakes he did. Firefighters often complained about feeling frustrated because of his management style. Because of this, morale was low and firefighters felt under-appreciated. You understand the importance of utilizing the talents, skills and abilities of others, and the advantages of getting others involved. Furthermore, you recognize that delegation is one of the most important aspects of time management. To help establish yourself as a confident officer who values input from others and promotes team unity, you decide to delegate effectively.

Scenario Questions

1. How would you go about delegating effectively?

2. What is the byproduct of effective delegation?

Supervision Scenario #3 - Answer Key

How do you delegate effectively?

1. Establish and maintain an environment that is favorable to delegating.

2. Select the right person for the job.

   ➤ Make sure the person has the talents, skills, ability and time to accomplish the task.

3. Assure the person understands the assignment.

   ➤ Make sure the person knows this is a forward step in his/her progress.

4. Keep an open door policy.

   ➤ Make yourself available to help those you delegate tasks too.

5. Be prepared to accept and deal with the consequences of that person's actions if they do not meet your Department's expectations.

6. Always reward performance.

What is the byproduct of effective delegation?

1. People see you as a stronger leader who has confidence in his team.

2. Increases morale throughout the department.

3. Produces greater overall efficiency.

4. Enables you to accomplish more.

Delegation; A Powerful Tool

BONUS!

The purpose of delegating is not to avoid work or unload difficult or tedious tasks to others. Effective delegation is an absolute necessity when it comes to a team's success. When you divide tasks, you multiply your chances of success. Failure to delegate will result in a failure to adequately develop your team. Through delegation, your team will grow in confidence. They – and your entire organization – will benefit in the long run.

You are a newly appointed Battalion Chief. Two Lieutenants come to you and tell you that a Captain's not performing up to your organizations standards. You ask them for specifics and they share information that leads you to believe the Captain's actions on the fire ground are in contrast to acceptable NFPA safety standards. Specifically, he provides inaccurate initial radio reports and does not call for enough resources at major incidents. He never requests additional Captains or the Battalion Chief at incidents that would require additional officers. Furthermore, the lieutenants are concerned because some of the firefighters were overheard complaining that they have "the worst officer on the job!"

Scenario Questions

1. What is your initial impression regarding the problem this officer is having based on the description above?

2. What information will you gather before the meeting?

3. How will you conduct this interview?

1. What is your initial impression?

The scenario narrative would lead one to believe the Captain is not familiar with or does not fully comply with the Incident Command System, NFPA standards, and department policies.

2. What information would you gather before the meeting?

- ❑ While meeting with the lieutenants, do the following:
  - ➤ Remind them of the Chain-of-Command, but be aware of the situation they are in.
  - ➤ Ask if they approached the Captain with this situation (find out what the results were).
  - ➤ Consider interviewing them separately to see if their stories match.
  - ➤ Assure them you will look into this matter and take appropriate actions to resolve the situation.
  - ➤ Document the meeting/interview.
- ❑ Get it in writing.
  - ➤ Have the Lieutenants provide written reports of what has been occurring.
- ❑ Gather facts – Review all pertinent info.
  - ➤ Personnel files of the lieutenants who are making the accusation.
  - ➤ Personnel file of the Captain in question.
  - ➤ Listen to recorded transmission and review any incident reports that can back up claims that the Captain doesn't call for appropriate resources, doesn't follow the ICS, provides inaccurate radio reports, and/or performs unsafe acts.
  - ➤ Determine the number of incidents where additional Captains or higher ranking officers should have been dispatched.
  - ➤ Investigate to see if this is a department wide problem, or isolated to this Captain.
  - ➤ Review department policies on the ICS, radio reports, alarm upgrades/resources,

NFPA safety standards, and any other department policies that may are being violated or ignored.

➢ Begin observing the Captains actions more closely.

3. **How will you conduct this interview?**

❑ Schedule a meeting – inform the Captain that he has the right to bring union representation.

❑ Conduct the Meeting.

➢ Begin by putting the member at ease.

➢ State the purpose of the meeting.

➢ Discuss positives about the individual before stating the problem at hand.

➢ Use facts to support the reason why this was brought to your attention and explain why it's inappropriate behavior.

❑ Get the individual's side of the story.

➢ Use open-ended questions to gather facts.

➢ Probe for answers.

➢ Show empathy.

❑ Look for an underlying problem.

➢ Circumstances in the individual's personal life may be affecting his/her behavior, especially since this behavior is uncharacteristic.

❑ Determine which of the 3U's you are dealing with (Unaware, Unable or Unwilling).

➢ The actions you take will be determined by which of the 3U's the individual falls under.

❑ Discuss progressive discipline - Verbal reprimand, written reprimand, suspension, fines, and termination.

➢ Explain what the next step may be if another issue arises.

❑ Develop a solution together and implement it.

➢ Take training and counseling into consideration:

➢ Training (NFPA safety standards, ICS, Initial radio reports, SOP's, organizational procedures, running schedule).

- ➤ Counseling (if there is an underlying problem, offer Counseling, Employee Assistance Programs, Critical Incident Stress Debriefing, etc.).

- ❑ Summarize – recap the key points to ensure clarity.

- ❑ Set a follow up meeting.

  - ➤ Stress that improvements need to be made within the specified time frame.

  - ➤ Open-door-policy.

- ❑ Inform the individual of the appeal process – Discuss Due process if they feel the actions taken to correct the situation is too harsh.

- ❑ Close on a positive note.

- ❑ Document and report.

  - ➤ Document in writing what occurred in the meeting and inform your superior of such.

  - ➤ Inform the firefighters who brought this to your attention that the issue has been addressed and corrective actions have been taken (do not give specifics).

- ❑ Monitor and evaluate the individual's progress until the next meeting.

- ❑ Get back to the Lieutenants to inform them this is being handled and to advise you of any additional concerns they have regarding this issue.

Underlining and Highlighting   BONUS!

When reading a scenario, underline or highlight the key points/problems that you must address in your answer. This will help you isolate the problem and focus on the key information presented, which will help you determine what actions need to be taken.

You are a newly appointed Battalion Chief. Two Lieutenants come to you and tell you that a Captain's <u>not performing up to your organizations standards</u>. You ask them for specifics and they share information that leads you to believe the Captain's <u>actions on the fire ground are in contrast to acceptable NFPA safety standards</u>. Specifically, he is provides <u>inaccurate initial radio reports</u> and <u>does not call for enough resources</u> at major incidents. He <u>never requests additional Captains or the Battalion Chief,</u> at incidents that would require additional officers. Furthermore, the lieutenants are concerned because some of the <u>firefighters were overheard complaining</u> that they have "the worst officer on the job!"

Supervision Scenario #5
Grievances

You are an officer who has recently been assigned to the grievance committee fter a senior member who held the position for 5 years retired. In your first week, you receive 3 grievances about a Deputy Chief who has been accused of showing favoritism to some and treating others unfairly. The DC overloads specific individuals with work, and often transfers people he clearly doesn't like. He disciplines some people for actions, while others – friends of his – get away with doing the same things.

Scenario Questions

1. What is the procedure for the person who will be handling grievances?

2. What are the negative effects of inconsistent discipline?

1. Procedure for the person who will be handling grievances?

1. Determine the responsibility to settle the grievance

2. Listen attentively as it is presented

3. Question the employee(s) to gather all the facts

4. Interview all parties involved

5. Keep adequate records

6. Analyze his/her alternatives

7. Decide who has the authority to act

8. Make a decision promptly

9. Explain his/her decision to the person representing the grievance

10. Follow Up to ensure improvement

11. Document

2. What are the effects of inconsistent discipline?

1. Poor Morale

2. Increased grievances

3. Lateness

4. Absenteeism

5. Increased legal fees

6. Poor Union relations

7. Freelancing

8. Non-compliance with rules and regulations

9. Increased injuries

10. Fighting among personnel

11. Confusion on what needs to be done or followed

Critical reading is the key to answering questions like this. The question was not "how would you handle the situation." The scenario really doesn't mean anything. The same questions could have been asked without any scenario.

Supervision Scenario #6
Freelancing

Dave is a 6 year veteran on his department. His 4 member Engine Company arrived on the scene of a warehouse fire and reported to you, the IC, for their assignment. They were the second Engine Company to arrive on the scene. You gave them the task of securing a secondary water supply from a nearby hydrant, and stretching a 2-1/2" hose line into the structure as a backup for the first line, which was already in operation. The driver, as always, stayed with the Engine to operate the pump and ensure the others received water. Dave's initial job was to connect the supply line to the hydrant so the engine could pump a continuous supply of water to the hose line. As he was doing this, the officer on his engine and another firefighter – the nozzle man – began to stretch the line into the smoky structure, where visibility was less than 10%. After securing the supply line, Dave was told to follow the line into the structure and meet up with his company so they could complete their task of backing up the first line and help extinguish the fire.

Ten minutes had gone by and Dave had not reunited with his crew. The officer radioed Dave to ask where his location was. There was no response. The Driver heard the transmission and radioed back that Dave had completed his first assignment (securing the water supply) then disappeared. He assumed Dave was somewhere inside the structure. The officer and the nozzle man, who were now running low on air, left their line and exited the building to find Dave. Because the officer could not locate Dave, he transmitted an urgent radio message stating that a firefighter was unaccounted for. This nearly halted all operations because the other firefighters heard the transmission and they heard you direct the rapid intervention crew to begin looking for the lost firefighter. Once outside of the structure, the officer found Dave working with a ladder company that was trying to force open some steel roll down doors on a loading dock for ventilation purposes. He did not realize his radio was turned off.

Dave said he had forgotten to turn on his radio and he was helping the ladder company because it looked like they could "use another pair of hands". The other members of the ladder company were so focused on their job that they did not realized they picked up an extra man. Dave was safe, but his freelancing caused many problems at this fire: The initial Engine Company was placed in a dangerous situation because they were still inside the structure without a back-up line. You had to change your overall tactics from fighting the fire, to a firefighter rescue mission. Other firefighters on the scene stopped what they were doing when they heard the urgent message. As a result, the fire grew larger and they ended up losing the warehouse. The entire mission was compromised and ultimately failed because of Dave's actions. When reviewing Dave's file, you discover that he had received a verbal reprimand 4 years ago from his previous superior officer for freelancing at a fire.

How would you conduct this subordinate interview, and what actions would you take to correct this problem?

❑ Gather facts – Review all pertinent info.

➤ Dave's personnel file.

➤ Personnel file of the officer on the ladder and all who are involved or who witnessed the incident at hand.

➤ Any incident reports where incidents concerning this individual were mentioned/documented.

❑ Get it in writing.

➤ Have Dave, his officer, and the officer on the ladder provide written reports of what occurred and why.

❑ Schedule a meeting.

➤ Dave should be provided with the option of having union representation.

❑ Conduct the Meeting.

➤ Begin by putting the firefighter at ease.

➤ State the purpose of the meeting.

➤ Discuss positives about the individual before stating the problem at hand.

➤ Use facts to support the reason why this was brought to your attention and explain why it's inappropriate behavior.

➤ When meeting with Dave, explain the following: The initial Engine Company was placed in a dangerous situation because they were still inside the structure without a back-up line. You, the IC, had to change your overall tactics from fighting the fire, to a firefighter rescue mission. Other firefighters on the scene stopped what they were doing when they heard the urgent message. As a result, the fire grew larger and they ended up losing the warehouse. The entire mission was compromised and ultimately failed because of his actions. Also mention that you are aware he has already received a verbal reprimand for freelancing at a fire several years ago.

❑ Get the individual's side of the story.

➤ Use open-ended questions to gather facts.

➤ Probe for answers.

- ➢ Show empathy.

- ❑ Look for an underlying problem.

  - ➢ Circumstances in the individual's personal life may be affecting his/her behavior, especially since this behavior is uncharacteristic.

- ❑ Determine which of the 3U's you are dealing with (Unaware, Unable or Unwilling).

  - ➢ The actions you take will be determined by which of the 3U's the individual falls under. Since Dave has already received a verbal reprimand, you can rule out Unaware. He is either Unable or Unwilling to take the correct actions.

- ❑ Discuss progressive discipline - Verbal reprimand, written reprimand, suspension, fines, and termination.

  - ➢ This will, at the very least, require a written reprimand. Remember: Any time you take disciplinary action, you should explain what the next step may be if another issue arises.

- ❑ Develop a solution together and implement it.

  - ➢ Dave should be required to complete training on the following: Training (NFPA safety standards, organizational procedures, fire ground operations training, and radio procedures training).

  - ➢ You should also encourage counseling to determine if there is an underlying problem he will not discuss with you: Counseling (Employee Assistance Programs, Critical Incident Stress Debriefing).

- ❑ Summarize – recap the key points and the solution to ensure clarity.

- ❑ Set a follow up meeting.

  - ➢ Stress that improvements need to be made within the specified time frame.

  - ➢ Open-door-policy.

- ❑ Inform the individual of the appeal process – Discuss Due process if they feel the actions taken to correct the situation is too harsh.

- ❑ Close on a positive note.

- ❑ Document and report – document in writing what occurred in the meeting and inform your superior of such. Also inform those who complained that corrective actions have been taken.

- ❑ Monitor and evaluate the individual's progress until the next meeting.

If Dave is **Unable** (Lacking mental or physical capability or efficiency; incompetent). You must give him a written reprimand at the very least. You will also follow the actions listed above.

If Dave is **Unwilling**, he is boldly resisting authority or having a defiant attitude. This is Insubordination and should be dealt with at a higher level. Dave will most likely be suspended and have to seek counseling, which is why you should cover your bases and offer counseling regardless.

Note: The process above should also be followed with the officer. You will want to meet him and determine if he was aware that this has previously been an issue with Dave. You will also want to know what corrective actions, if any, were taken. In short, you must determine if the officer is doing his job? If so, the problem lies with Dave. If not, the officer should be required to participate in mandatory training and receive an verbal reprimand.

The same can be said about the officer on the ladder. You need to find out why he was not aware that he picked up an extra member, especially after an urgent radio transmission and RIC deployment. Determine if he was Unaware of the situation and provide training.

Long Narratives

BONUS!

Don't let an unusually long scenario narrative throw you off. Remember to underline key points the first time you read the scenario so if you have to go back, it will be easy to locate them. Also remember, the key to success is in your answer. When you package your answer correctly, you will not be intimidated by a detailed scenario.

Administrative Scenarios

Administrative Scenario #1
Develop a Community Program

Your department has been receiving negative press lately due to a fight involving two members. Several designated department officials have been addressing the incident. In an effort to create a more positive public image, and take the attention away from the incident, your department head decided this would be a great time to initiate a public relations program that is designed to encourage young children to read. A group of firefighters started working on creating this program two years ago, but never moved past the initial planning stage. You are a newly appointed officer and you have been given the important assignment of creating this program.

Scenario Question

How would you go about developing and implementing this type of program?

*Administrative Scenario Tip: One of the best ways to tackle administrative scenarios is by following the PRDIE format outlined in the books Step Up and Lead and Fireground Operational Guides.

One way to tackle this assignment is by utilizing the 5 step P.R.D.I.E format outlined in the book Step Up and Lead. The letters stand for <u>P</u>lan, <u>R</u>esearch, <u>D</u>evelop, <u>I</u>mplement, and <u>E</u>valuate).

1. The PLANNING stage includes the following actions:

   - ☐ Meet with the chief to discuss the departments needs and identify any problems.

   - ☐ Form a committee with motivated and knowledgeable firefighters and subject matter experts.

     - ➤ For example, coworkers who have children or teaching experience, Legal experts, Representatives of groups that will be affected by the program such as local teachers or board of education members, anyone else who could provide valuable insights. Include members from the previous committee who began working on this project a year ago.

   - ☐ Review all the information gathered by the previous committee members and use that information if it is still relevant and helpful.

   - ☐ Assign a committee manager to facilitate coordination between all committee members. (Since you have been assigned this task, you should assume this role).

   - ☐ Schedule a committee meeting. At the meeting, do the following:

     - ➤ Clearly define specific goals and objectives (short, mid-range, long-term).

     - ➤ Prescribe a course of action by delegating tasks to each committee member (early tasks are mainly fact-finding and research-oriented). Examples of things to research are provided in the research stage.

     - ➤ Motivate your committee members by offering rewards for their time and efforts.

     - ➤ Make sure members write down their assignments so that there is no confusion about who is assigned which task.

     - ➤ Schedule your next meeting, and make sure all members know when they should have their tasks accomplished.

   - ☐ Document the details of the meeting in writing and inform your superior of the direction that has been set; then move to the next step. (These two actions should be taken after each of the five steps in the process).

2. The RESEARCH stage includes the following actions:

☐ Research all pertinent facts that were assigned at the committee meeting. Be sure to check:

➢ NFPA standards

➢ Other fire departments that may have a similar program or SOP on the topic

➢ County Office of Emergency Management (OEM)

➢ Internet search

➢ Subject matter experts

➢ Product manufacturers

➢ Successful reading programs such as "Read Across America."

☐ Research the costs that will be involved in implementing the program, and ensure you have or can raise any necessary funding. Costs can include:

➢ Staff training

➢ Yearly maintenance

➢ Tools and supplies

☐ Research ways to offset costs

➢ Available grants

➢ Pooling resources such as training facilities, equipment, money with other fire departments

☐ Identify your target audience (for example, fourth graders).

☐ Research the types of books you will want to purchase for the program.

☐ Identify which individuals will be interacting directly with the children.

☐ Document the details of your research, inform your superior of your findings, and move to next step.

3. The DEVELOPMENT stage includes the following actions:

☐ Bring the committee back to discuss their findings.

☐ Discuss all researched information.

☐ Begin formatting and writing the procedure (SOP) that covers the program. An effective SOP includes the following sections: Purpose, Scope, Responsibility, and Procedures or Guidelines. At the very least, you want to make sure the document you develop includes:

> ➢ The reason why the document was develop.

> ➢ An outline of when and where this program and procedure shall apply.

> ➢ A list of each person (or job title) who shares the responsibility of implementing the program.

> ➢ Specific guidelines, techniques, and methods that each participant must follow in order to ensure success.

> ➢ A mechanism for evaluating the effectiveness of the program. (This will be discussed more in Steps 4 and 5.).

☐ When complete, have each committee member review the new document for technical accuracy and grammatical errors.

☐ Document the details up to this point, inform your superior, and move to next step.

4. The IMPLEMENTATION stage includes the following actions:

☐ Evaluate the program's effectiveness before fully implementing.

☐ Practice the techniques outlined in your new SOP.

☐ Choose one school/class to test your pilot program. Use this opportunity to work out any kinks.

☐ Utilize as many people as needed in order to assess whether you feel your efforts enabled you to meet your goals.

☐ Make revisions to the document, if needed.

☐ Meet with the chief to discuss your final version.

☐ Begin the implementation process.

☐ Send out the SOP with a special notice and implementation date.

☐ Require all personnel to review the SOP.

- [ ] Set up training evolutions to practice the procedures outlined in the SOP.

- [ ] Conduct tabletop drills to discuss the procedures outlined in the SOP.

- [ ] Announce the program to the public. (Remember this is a public relations program) The following are some possible ways to inform the public of your new program:

  - ➤ Local newspapers (contact the publisher or your community reporter)

  - ➤ Local television programs (contact the program's producer)

  - ➤ Brochures mailed to all local residents (there may be significant costs with this method)

  - ➤ The Internet (post the program on your organization or community website; develop a page on popular social networks)

  - ➤ Community notification channels on television and school handouts that teachers can give their students to bring home

  - ➤ You can choose one or all of these methods. Your best option would be a combination of those listed, otherwise known as a multimedia campaign. Whichever method you choose, be sure to include all essential information, such as program locations, times, and dates.

- [ ] Implement the program and monitor its effectiveness by using the mechanism for evaluating that the committee developed in the developmental stage.

- [ ] Add any recurring expenses to the department's line-item budget.

- [ ] Document the details up to this point, inform your superior, and move to next step.

5. The EVALUATION stage includes the following actions:

  - [ ] Evaluate the effectiveness of the program; ask questions such as:

    - ➤ Is this program still an effective way to meet our goals and objectives?

    - ➤ Have we improved our relations with the public we serve?

    - ➤ Have we accomplished our goals of encouraging young children to read?

    - ➤ Does this program contradict acceptable methods or public relations standards that are followed throughout the fire service?

    - ➤ Have we met our needs?

> ➤ Do our members feel this SOP is the "one best way"?

> ➤ How can we improve?

☐ Compare pre-program results with post-program results to see if your deptartment has improved.

☐ After you review and evaluate the effectiveness of the program, revise if needed.

☐ Review your program at least once a year. Cycle the programs through the planning stage often to ensure it continues to meet your goals and objectives.

☐ Document, inform your superior, and recycle back into the planning stage when needed.

☐ Thank and reward the committee members for their service.

Types of Programs

There is no shortage of the types of community service programs a fire service organization can develop. Below is a list. If you are presented with a scenario to create any of the programs below, consider using the P.R.D.I.E. format as a guideline.

- Basic First Aid
- BBQ grills and summer safety tips
- Bicycle safety
- Carbon monoxide safety tips
- Children's car seats
- CPR and AED
- Fire drills (public assemblies)
- Fire extinguishers
- Fire works
- Hazardous materials (storage and use)
- Home escape planning, E.D.I.T.H.
- Home fire prevention
- Home heating dangers/emergencies
- How to use Emergency 911
- Natural disaster preparation
- Neighborhood watch programs
- Smoke detectors
- Stop, drop and roll
- Winter Safety Tips

Administrative Scenario #2
Develop a Standard Operating Procedure

Photo By: Ron Jeffers

The head of your department brought you into his office and said he has been impressed with your work ethic and ability to "get things done." He considers you to be a forward thinker and problem solver. Because of this, he has chosen to delegate an important assignment to you.

You have been assigned the task of developing a Standard Operating Procedure for a new challenge in your community – Fires in high rises under construction.

Scenario Questions

1. How would you go about developing and implementing this SOP?

2. What are the advantages and disadvantages of SOP's?

Question #1:
How would you go about developing and implementing this SOP?

1. The PLANNING stage includes the following actions:

   ☐ Meet with the chief to discuss department needs and identify problems.

   ☐ Form a committee with motivated and knowledgeable firefighters and subject matter experts.

   > ➤ For example, high-rise construction contractors, building owners and managers, legal experts, other department officials who have knowledge on the subject, and anyone else who could provide valuable insights.

   ☐ Assign a committee manager to facilitate coordination between all committee members. (Since you have been assigned this task, you should assume this role).

   ☐ Schedule a committee meeting. At the meeting, do the following:

   > ➤ Clearly define specific goals and objectives (short, mid-range, long-term).

   > ➤ Prescribe a course of action by delegating tasks to each committee member (early tasks are mainly fact-finding and research-oriented). Examples of things to research are provided in the research stage.

   > ➤ Motivate your committee members by offering rewards for their time and efforts.

   > ➤ Make sure members write down their assignments so that there is no confusion about who is assigned which task.

   > ➤ Schedule your next meeting, and make sure all members know when they should have their tasks accomplished.

   ☐ Document the details of that meeting in writing and inform your superior of the direction that has been set; then move to the next step. (These two actions should be taken after each of the five steps in the process).

2. The RESEARCH stage includes the following actions:

- [ ] Research all pertinent facts that were assigned at the committee meeting. Be sure to check:

  - ➢ NFPA standards

  - ➢ Other fire departments that may have a similar program or SOP on the topic

  - ➢ County Office of Emergency Management (OEM)

  - ➢ Internet search

  - ➢ Subject matter experts

- [ ] Research the costs that will be involved in implementing the SOP, and ensure you have or can raise any necessary funding. Costs can include:

  - ➢ Staff training

  - ➢ Yearly maintenance

  - ➢ Essential tools and equipment

- [ ] Research ways to offset costs.

  - ➢ Available grants

  - ➢ Pool recourses such as training facilities, equipment, money with other fire departments

- [ ] Document the details of your research, inform your superior of your findings, and move to next step.

3. The DEVELOPMENT stage includes the following actions:

   - [ ] Bring the committee back to discuss their findings.

   - [ ] Discuss all researched information.

   - [ ] Begin formatting and writing the procedure (SOP). At the very least, an effective SOP includes the following sections:

     - ➢ Purpose: The reason the document was developed

     - ➢ Scope: An outline of when and where this program and procedure shall apply

     - ➢ Responsibility: A list of each person (or job title) who shares the responsibility of implementing the program

     - ➢ Procedures or Guidelines: Specific guidelines, techniques and methods that each participant must follow in order to ensure success

- [ ] The Procedures/Guidelines section should address the following (at a minimum):

  - ➢ Construction information, including elevator, stairwell and HVAC system info.

  - ➢ Copy of the floor plan and utility controls

  - ➢ Contact names and number for building representative/manager

  - ➢ Engine, Ladder and Rescue company Operations

  - ➢ Incident Command system – structure

  - ➢ Running schedule and list of resources that can be utilized

- [ ] Identify a mechanism for evaluating the effectiveness of the program. (This will be discussed more in Steps 4 and 5.).

- [ ] When complete, have each committee member review the new document for technical accuracy and grammatical errors.

- [ ] Document the details up to this point, inform your superior, and move to next step.

4. The IMPLEMENTATION stage includes the following actions:

   - [ ] Evaluate the program's effectiveness before fully implementing.

   - [ ] Practice the techniques outlined in your new SOP.

   - [ ] Choose one similar structure to test your pilot program. Use this opportunity to work out any kinks.

   - [ ] Utilize as many people as needed in order to assess whether you feel your efforts enabled you to meet your goals.

   - [ ] Make revisions to the document, if needed.

   - [ ] Meet with your department head to discuss your final version.

   - [ ] Begin the implementation process.

   - [ ] Send out the SOP with a special notice and implementation date.

   - [ ] Require all personnel to review the SOP.

   - [ ] Set up training evolutions to practice the procedures outlined in the SOP.

   - [ ] Conduct tabletop drills to discuss the procedures outlined in the SOP.

- [ ] Add any recurring expenses to the department's line-item budget.

- [ ] Document the details up to this point, inform your superior, and move to next step.

5. The EVALUATION stage includes the following actions:

- [ ] Evaluate the effectiveness of the SOP; ask questions such as:

  - ➤ Is this SOP still an effective way to meet our goals and objectives?

  - ➤ Have we improved in our ability to serve the public?

  - ➤ Have we accomplished our goals of improving operations at these fires?

  - ➤ Does this program contradict acceptable methods or public relations standards that are followed throughout the fire service?

  - ➤ Have we met our needs?

  - ➤ Do our members feel this SOP is the "one best way"?

  - ➤ How can we improve?

- [ ] Compare pre-SOP results with post-SOP results to see if your department has improved.

- [ ] After you review and evaluate the effectiveness of the SOP, revise if needed.

- [ ] Review the document at least once a year to ensure it continues to meet your goals and objectives.

- [ ] Document, inform your superior, and recycle back into the planning stage when needed.

- [ ] Thank and reward the committee members for their service.

Question #2:
What are the advantages and disadvantages of SOP's?

Advantages:

1. Facilitates Delegation

2. Allow for consistency

3. Enable leaders to manage more effectively

4. Are based on the "one best way"

Disadvantages:

1. Lack flexibility

2. Tend to become obsolete

3. Tend to stifle initiative

4. Take time and money to develop

You have recently discovered that many of your department members have not been properly trained on how to call a mayday. By studying previous events, you also discover that Mayday radio transmissions will often occur at the worst possible time iuring an incident. An escalating or deteriorating incident, accompanying radio traffic, combined with the anxiety and confusion that could come from hearing either a whole or partial emergency radio transmission requires that Mayday radio transmissions are clear and contain an easily recognizable set of parameters. Your department does not have a standard operating procedure on Mayday radio transmission. Your Chief would like an SOP developed within 10 days. You, as a newly promoted officer, have been assigned the task of providing the following information:

1. Definition of what a Mayday is (as far as our job is concerned),

2. Mayday parameters (When a FF/officer should transmit a mayday),

3. The minimum amount of info a FF should provide when calling a Mayday, and

4. How the IC should acknowledge and manage a mayday.

Scenario Question

What information would you provide in each of those four categories?

1. Definition of the word Mayday

Use of the word Mayday identifies that a firefighter/fire officer has become lost, trapped, or seriously injured, or has exhausted his or her breathing air at the scene of an emergency incident. Specifically, a firefighter is in need of immediate help.

2. Mayday parameters: a firefighter/fire officer will transmit a Mayday if any of the following conditions exist:

   ☐ If you become lost, trapped, or have sustained a serious or life-threatening injury.

   ☐ When a serious or life-threatening injury to another member has occurred.

   ☐ If you discover a lost, trapped, seriously injured or unconscious firefighter.

   ☐ If you become tangled or pinned and are unable to free yourself after the first attempt.

   ☐ If your low-air alarm is activated and you are unable to find an exit.

   ☐ If there is zero visibility, you have no contact with a hose line or search rope, and you do not know where the exit is.

   ☐ If your primary exit is blocked by fire or collapse, and you cannot locate an immediate secondary exit.

   ☐ If you fall through a floor, roof, staircase, or down a shaft.

   ☐ If you are caught in a rollover condition and cannot find an exit.

   ☐ Any life or death situations where you need immediate assistance.

3. Mayday radio procedure should include the following:

   ☐ Activate the emergency identifier button (EIB) on your portable radio if you cannot talk.

   ☐ In an attempt to send out as much useful information as possible in the shortest amount of time, remember M3-W3

      ➤ M: Mayday (to be announced three times)

> Who: Identify who you are (your radio designation, ex: Ladder 10, Engine 20)

> What: Give your situation (lost, trapped, injured, and so on)

> Where: Give your location, such as floor, side, other (ex: third floor, side C)

☐ The member transmitting the Mayday must pause after each message and then repeat the message until acknowledged by the incident commander.

☐ The fire department dispatch center must relay any Mayday messages that are not immediately acknowledged by the incident commander.

☐ Members shall also activate their personal alert safety system/device (PASS) in between each message and after acknowledged. IMPORTANT: If the PASS device remains activated during the transmission of the Mayday message, it will cause significant background noise and make the message unreadable.

☐ LUNAR is an acronym to help the IC remember what information to obtain. LUNAR stands for:

> L – Location (where in building, or your assignment)

> U – Unit (Apparatus)

> N – Name (Your name vs. your designated position)

> A – Air (remaining supply)

> R – Resources (what think you need)

4. Mayday radio acknowledgment and management should include the following:

Acknowledgment:

☐ Mayday transmissions take priority over all other transmissions, including urgent messages. There are no exceptions!

☐ When a Mayday transmission has occurred, the IC must attempt to clear the air of all other radio traffic and establish contact with the lost, trapped, or injured member(s).

☐ After contact is established, the IC should attempt to obtain more specific information that may assist in the rescue attempt if it proves necessary.

☐ The requesting of other or more specific information will be determined by the amount of information originally transmitted in the Mayday. Information requests may include but are not limited to:

- ➤ Can you tell us the best/closest access route to you?

- ➤ Can you hear a hose stream or saw operating nearby?

- ➤ Are you near a stairway, shaft way, wall, or other building feature?

- ➤ Can you give the condition of the injured member(s)?

- ➤ What tools and equipment will we need?

- ➤ Can you offer any other useful information?

Managing the Mayday:

IC's must maintain control and continuity of the incident by any means available to them. Options that are available to all Incident Commanders include but are not limited to:

- ☐ Using multiple rapid intervention crews for deployment and replacement.

- ☐ Transmitting of an additional alarm(s).

- ☐ Designating a separate radio frequency for the fire operation.

- ☐ Conducting a personnel accountability roll call to determine who/how many are missing.

- ☐ Collecting accountability tags and riding lists to determine who and how many are missing.

- ☐ Reviewing crew assignment and command tracking to ID members last assigned location.

- ☐ Establishing and supporting a rescue group/operation within your incident management.

- ☐ Verifying that fire suppression operations are continuing.

- ☐ Removing all nonessential personnel.

- ☐ Eliminating freelancing and establish control.

- ☐ Requesting any additional resources and equipment that may be needed.

Urgent Radio Guidelines

BONUS!

Mayday and Urgent radio transmission are different. The word "urgent" should be used to identify that a potential life-threatening situation has developed that may affect firefighter and civilian safety. The process of calling an urgent message is the same as calling a mayday and can be summed up with U3-W3

Urgent, Urgent, Urgent – Who, What, Where

Photo By: Brett Dzadik

You are a newly appointed officer. Your superior has just informed you that within 3 months, a new highway that will run along the edge of your community is going to be constructed. Within the past year, two local warehouse facilities began storing and transporting hazardous materials. Your department does not have an SOP outlining response guidelines for hazardous material incidents. The closest hazardous material response team is nearly 20 minutes away.

Scenario Questions

You have been given the assignment of providing information on the 8 Step Plan for tactical management of hazmat incidents. What are the 8 steps and what information would you include under each category?

The 8 step plan for tactical management of hazmat incidents: SHIP IRDT

1. Site Management and Control

   ☐ Establish a security perimeter by isolating and denying entry to the area/building. Give significant consideration in using the police in assisting you with this responsibility.

   ☐ Establish control zones for the incident site: Hot, Warm, and Cold.

   ☐ Assure a safe approach for incoming resources.

   ☐ Establish staging (Uphill and Upwind) as a method of controlling arriving resources.

   ☐ Implement public protective actions by evacuating, protecting in place, or implementing a combination of the two.

2. Identify the Problem

   ☐ Determine building occupancy and location; the name on the front of the building may tip you off to specific concerns.

   ☐ Container shape can indicate pressurized vs. non-pressurized containers.

   ☐ Marking and colors as well as placards and labels can provide product information and mitigation options to first responders.

   ☐ Shipping papers, facility documents and Material Safety Data Sheets (MSDS) will provide product information.

   ☐ Monitoring and detection equipment can identify the presence of a product.

   ☐ Your senses can help identify the problem. Pay attention to what you see, hear, and smell.

3. Hazard & Risk Evaluation

   Hazard: This is defined as the Physical and Chemical Properties of a material.

Risk: The intangible, undesirable probability of suffering harm or loss.

Gather hazard data from the following:

- [ ] Reference materials (Department of Transportation Emergency Response Guidebook – DOT/ERG)

- [ ] Technical information centers such as CHEM-TREC

- [ ] Hazardous material databases

- [ ] Right-to-know information; Material Safety Data Sheets (MSDS)

- [ ] Monitoring instrument

- [ ] Determine the extent of damage to container

- [ ] Predict the likely behavior of the released material and containers

- [ ] Analyze HAZARDS & RISKS to determine the safest and most effective Incident Action Plan

4. <u>Select proper PPE & Protective Clothing</u>

- [ ] Is HazMat gear/equipment needed, or will firefighting gear/respiratory protection be enough?

5. <u>Information Management and Resource Coordination (ICS)</u> - U SIL FLOP SR

Use the Incident Command System at ALL structure fires, regardless of size/scope. Assignments may include the following command and support staff members:

- [ ] Unified Command Staff (should be considered)

- [ ] Safety officer (to ensure a safe working environment)

- [ ] Information officer (to communicate with the media and concerned citizens)

- [ ] Liaison officer (to communicate with other agencies and ensure smooth operations)

- [ ] Finance officer/section

- [ ] Logistics officer/section

- [ ] Operations officer

- [ ] Planning officer/section

□ Staging officer

□ Rehabilitation officer

Also: Establish Divisions/Groups early to enhance communications and improve accountability. Don't wait to be overwhelmed with your span of control. (Example: Rescue = Rescue Group; Spill containment = Containment Group, etc.) Assigning division supervisors will help you manage your span of control. Less people trying to communicate with you directly ensures a greater level of effectiveness.

6. <u>Implement Response Objectives</u>

The level of risk will determine how and if you deploy your hazardous material and mitigation team(s). From the hazard and risk evaluation, start off with determining:

□ Offensive operations—Rescue, containment, etc

□ Defensive operations—confinement

□ Combination of both, or

□ Non-intervention—it takes its natural course

From here you can identify more specific actions and objectives from the challenges presented.

□ Rescue—for example, extricate the driver

□ Spill control/confinement—for example, dike and dam the spill

□ Leak control/containment—for example, plug, patch, shut valves

□ Fire control—for example, eliminate ignition sources, de-energize and stabilize the vehicle, suppress vapors with foam, extinguish the fire

□ Public protective actions—for example, evacuate or protect in place the occupants of a nursing home or day care center

*Assign Engine and Ladder companies (see below), especially if there is a fire at the incident:

Engine Company Ops: Here is a list of the key points you will want to address with engine co's.

- ☐ Position apparatus
- ☐ Establish a primary (and secondary) water supply
- ☐ Initiate attack
- ☐ Choose appropriate size hose lines
- ☐ Advance and position hose lines
- ☐ Locate, Confine, & Extinguish (LCE) the fire
- ☐ Protect exposures
- ☐ Supply auxiliary appliances
- ☐ Utilize Thermal Imaging Cameras (TIC)
- ☐ Coordinate with ladder companies (and other companies on scene)
- ☐ Provide periodic progress reports

Ladder Company Ops: LOVERS-UPS (TIC-COP)

- ☐ Position apparatus
- ☐ Raise and position Ladders
- ☐ Force Entry
- ☐ Primary and Secondary Search
- ☐ Rescue Operations
- ☐ Ventilation (Horizontal, Vertical, Outside Vent Member)
- ☐ Utility control
- ☐ Salvage
- ☐ Overhaul
- ☐ Utilize TIC's
- ☐ Coordinate with engine companies (and other companies on scene)
- ☐ Provide periodic Progress reports

7. <u>Decontamination</u>

☐ Assume anything coming out of the "HOT ZONE" is exposed and potentially contaminated

☐ Establish a Decon site/officer and Group.

☐ Conduct DECON and medical monitoring for all members

8. <u>Terminate the Incident</u>

☐ <u>T</u>ransfer to another officer or Terminate completely

☐ Conduct an <u>I</u>ncident Debriefing

☐ Schedule a <u>P</u>IA (Post Incident Analysis)

☐ <u>D</u>ocument the incident, complete reports

☐ CISD (request if needed)

Practice the 8-Step Plan

BONUS!

In this scenario, you weren't asked to develop the SOP. You were given a specific assignment, which was to provide information that will be added into the SOP. Furthermore, this specific scenario was created to help you practice the 8-Step plan, which will get a fire officer through any simulated or real life haz-mat scenario. Never underestimate the importance of repetition.

You have been assigned to the training division and are assuming the position of training officer. You have been asked by the head of your department to develop a committee of subordinates and create a program on home fire prevention. After choosing your committee members, you schedule a meeting and discuss your options before deciding upon a direction to go with this new program. As program manager, you delegate tasks to each individual who is part of the committee because you understand that delegation is one of the most important aspects of effective time management.

Scenario Questions

1. In order for supervisors to delegate effectively, what must they do?

2. Explain what happens when you try to do it all yourself compared to the byproducts of effective delegation.

1. In order for supervisors to delegate effectively they must:

   ☐ Feel secure about their own position.

   ☐ Establish an environment that's favorable to delegating.

   ☐ Understand the importance of utilizing the talents, skills and abilities of those around them so you can choose the right person for the job.

   ☐ Assure the person accepting the assignment sees it as a forward step in his/her career.

   ☐ Keep an open door policy and be available to give subordinate assistance.

   ☐ Ensure the person accepting the assignment has the ability, time, and resources to do the job and fully understands the assignment they are given.

   ☐ Be prepared to accept & deal with the consequences of the subordinates actions.

   ☐ Always reward performance.

2. What happens when you try to do it all yourself?

   1. You only have your own personal input.

   2. You fail to develop your team.

   3. You create unnecessary stress & burden.

   4. You develop health issues.

Byproducts of Effective Delegation

BONUS!

1. People see you as a stronger leader who has confidence in his team.

2. Increases morale throughout the Dept.

3. Produces greater overall efficiency.

4. Enables you to accomplish more.

During an annual run survey review of the previous year, you discover that one-sixth of your departments fire responses were vehicle fires on Highways. These fires accounted for more civilian injuries and deaths and more firefighter injuries than all your other responses combined. Your members are well trained in fighting vehicle fires and performing extrications; however, you identified a serious problem concerning these highway fires. Your members are not setting up sufficient safe work zones to protect themselves. Because of this, you initiate the creation of an operational guide for fighting fires on highways.

Scenario Question

What actions would you list to help ensure the safety of your members who are approaching and working at fires on highways?

Administrative Scenario #6 - Answer Key

1. Approach slowly, conduct a thorough Size up and Exit your vehicle with caution.

   - Lookout for fuel spills, victims, downed wires and other hazards that may have resulted from or caused the accident.

   - Beware of curious drivers (rubberneckers).

   - Make sure traffic has stopped, but never assume traffic has ceased. Respect reaction delays and stopping distances.

Speed	Stopping distance	Reaction delay	Total stopping distance
20 mph	20-feet	22 feet	42 feet
40 mph	80-feet	44 feet	124 feet
60 mph	180-feet	66 feet	246 feet

2. Request Additional Resources

   - Call for additional apparatus for firefighter safety.

   - Rescue/Extrication Company, if there is an entrapment.

   - Additional Companies if exposures are threatened.

   - Police for traffic/crowd control; Firefighters should divert or stop traffic until Police arrive on scene.

   - EMS (ambulances) for patient treatment.

   - Mobile Water Supply Vehicle (or additional Engine Companies for Relay or Shuttle operations) if needed for water supply.

   - Foam Tender (Call early, if the situation dictates).

3. Protect Firefighters by Establishing a Safe Zone

   - Use warning devices such as cones, reflectors, hand lights, apparatus lights and/or flares (If you use flares, be aware of flammable liquid runoff).

   - Use apparatus to block the lane closest to the vehicle fire.

   - Position your Command Vehicle to alert oncoming traffic and protect firefighters.

- Position apparatus defensively, in between moving vehicles and firefighters.

- Have additional apparatus respond to block traffic (at least two apparatus).

  ➢ The first Engine should be used as a blocker.

  ➢ The Second apparatus should be used as a shadow vehicle for additional protection.

- Position pump panels away from traffic.

- Conduct operations from the shoulder. Keep firefighters away from moving traffic.

4. Dress in Reflective gear. IC's and those who are not extinguishing the fire should always don a reflective vest.

Vehicle Fires on Highways

BONUS!

Approximately one-sixth of all annual fire responses are vehicle fires on highways. These fires account for more than 2,000 civilian injuries and deaths and nearly 1,100 firefighter injuries each year.

Oral Presentations and the Fire Service Examination Process

When it comes to promotional testing, having a vast knowledge base is essential, but without good communication skills the oral portion of a promotional exam or interview can be daunting. The cold, hard fact is that a firefighter with an extensive knowledge base but poor communication skills is at the mercy of the lesser educated firefighter who knows how to communicate a message in a clear and confident manner. When it comes to the oral examination, you must know how the system works and you must "play the game" by their rules. With that in mind, this document will help you prepare for you oral examination by providing you with key pointers that will help you maximize your score and make a good impression.

Breaking it down

First, you must understand that this process varies from state to state and by the rank you are testing for. You may have already taken a written exam and now you are entering a room where you are expected to read a scenario and give a ten or fifteen-minute presentation to a video camera. The recording will be reviewed and graded by a team of assessors at a later date. Perhaps you are testing for a chief officer position and you have been given a specific amount of time to review your scenario(s). Now you have to enter a room where you are going to be graded by a panel of your peers on your knowledge base and your oral presentation skills. Because the system varies from state to state, it is impossible for this document to target your specific testing process; however, there are "absolutes" when it comes to making a good presentation regardless of the rank you are testing for.

This document was developed to assist you in maximizing your score. Applying this information in conjunction with a strong knowledge base on the subject you are being tested on will give you the edge over your competition.

Communication Skills - making your oral presentations

A good study group and strong work ethic will arm you with the knowledge base we mentioned earlier. Along the way, you will come across tips on what to say and how to say it. Those tips may help you score higher on the test, but if you are serious about elevating yourself above the crowd and overcoming handicaps – like seniority or lack of experience – then you should be aware of the subconscious aspects of an oral presentation.

Subconscious aspects

When you are being assessed, it's not only your words that are scrutinized during an oral presentation; it's also your voice, your body language, and your appearance. Here is a quick breakdown of all three:

- Your voice - How you say something is as important as what you are saying. It's essential to present yourself as a confident–but not cocky, fire officer.

- Body language – This is a subject in its own right and a topic in which much has been written and said. In essence, your body movements express what your attitudes and thoughts really are. On the pages that follow, you will find tips on body language that you will find helpful.

- Appearance - First impressions influence people's attitudes toward you. Dressing appropriately for an occasion is critical. Some feel a suit is sufficient. Others believe your organizations dress uniform is the right choice. This has been a debate since the beginning of the testing cycle and I will provide my thoughts later in this document.

As with most personal skills, oral communication cannot be taught. Instructors can only point the way. As always, practice is essential to improve your skills and prepare to provide the best presentation you are capable of.

Preparation

Unless your last name is Hilton or Trump, success in any arena is not handed to a person on a silver platter. If you are serious about advancing your career and achieving higher ranks, the first thing you should do is to identify the reason or reasons WHY. In other words, what is the driving reason why you want to be a lieutenant, captain, battalion chief, deputy chief or whatever other position you desire? Your answer may be self-fulfillment, money, or simply because you know you would make a great officer. The important thing is that you have an answer – a reason. Without it, you will likely find a reason why preparing, or studying, is more of an inconvenience than a necessity.

Once you identify that reason, and have dedicated the necessary time reading the appropriate books (which is essential) and attending educational seminars (which is highly recommended), you can focus on three key things:

1. Practice,

2. Practice, and

3. Practice.

You don't want to make the mistake that so many others do, which is showing up on test day unprepared. If the first time you give an oral presentation is on the day of your test, you are not playing your hand well. Instead of "winging it", prepare as if your career depends on it. Two of the Chief Officers on our staff at FireOpsOnline admit to taking a mock test every day for 30-45 days prior to their exams. On the day of the test, they were fully prepared to read their scenarios and organize their notes in the allotted time frame. They were also prepared to enter the assessment room and give their oral presentation (4 scenarios: fire, non-fire, administrative and subordinate interview) to a panel of assessors. Of course, test day was game day, but for these would-be officers, game day was just like any other day because they practiced everyday like it was game day.

What exactly are you preparing? In the shortest possible answer, you are preparing to hit the main points that will provide you with the best possible score, and you are preparing to deliver your message in the most professional manner possible.

As you study, you will begin to carefully and logically prepare the structure of your presentation. Although the questions or scenarios will vary, your approach to answering them will remain the same. For example, when studying for the Deputy Chief exam, you should prepare for every scenario imaginable. One likely scenario will be to develop a program of some sort (such as a community awareness program that provides seniors with holiday safety tips for the winter months). The specifics of the program will change, but the format you use to develop it will revolve around the same steps. So, if the question is 'What steps you would take to develop a fire safety awareness program for your community?' you would immediately refer to your format. An example would be the 5 step approach known as P-R-D-I-E (that format is thoroughly covered in this book). Whether you are asked to develop an SOP for your department or a community awareness program, that format works.

Know your testing process

Be sure to familiarize yourself with your testing process so that you know what types of questions will be asked. For example, if testing for the position of Deputy Chief, you may have to study and prepare formats for the four scenario categories we mentioned earlier: (fire, non-fire, administrative and subordinate interview). I can't stress enough that, although the details of each scenario would change, the formats and approach to those scenarios would not. Memorizing acronyms and formats will provide you with a great advantage. Doing so will enable you to focus less on what you need to say and more on your communication skills and oral presentation.

Here's how you develop your own formats: Write your formats out in rough drafts, like the first draft of a written report or article. When you review the draft, you will find things that are repetitive, irrelevant or superfluous – delete them. Your format will look more like notes at this point. Keep scaling it down and practicing until your answer is consistent and flows smoothly, and the key points you need to hit are reduced to single words (see the fire scenario format on the next page).

It is unwise to try and write your format (or notes) in detail on exam day. First, you will not have enough time to do so; and second, even if you did, chances are you would not be able to locate the thing you want to say amongst all the other text you jot down on your note pad. Instead, through proper preparation you should know what you want to say, and through properly bulleted note organizing, you should be able to glance down at your notes, and locate the key word(s) or phrase(s) you are looking for.

Once you scale your format down to the bare minimum, you will be able to review your scenario, write out your format, and fill in the necessary areas with scenario specific content. The point is simple, some people get so caught up in trying to address all the challenges that the scenario presents that they forget to do simple things, like call for a second alarm or secure a water supply.

When your format is down on paper, and you are ready to give your oral presentation, never read straight from your notes. Simply glance down, find the word that represents what you need to address, and thoroughly address that point.

In Fireground Operational Guides, Deputy Chief's Mike Terpak and I provide readers with a format to follow when writing a narrative for a structure fire report. That is the same format you can use when testing for the Battalion or Deputy Chief position and you are presented with a scenario where you need to tackle a structure fire – from the initial call until reports are completed.

On the following page is a sample format to a fire scenario (Refer to Fireground Operational Guides for additional information on this particular format). Note: This one is slightly different from the one you learned in this book, but look at how the end result is the same... you cover all the essential points.

Your universal format, when written down, may look something like this:

En route (monitor radio, review pre-plans, multi-sided view)

Est. Command (location of CP)

Size-Up: *COAL TWAS WEALTH

IRR: (radio report, *CAR - Conditions, Actions, Resources)

Resources: *2U PERS WAR +

Implement Incident Command System:

Determine Strategy and Tactics:

Engine Company Ops: *WASS POCC

Ladder Company Ops: *AL VES SCOUP

Under control: *SS PIDS CO

Terminate command: *TIP RC

* refer to the book Fireground Operational Guides for acronym descriptions.

The benefit of a good format is obvious... regardless of the specifics in you fire scenario, a good format will remind you to address the ESSENTIALS of ALL fire-ground activities. For example, in the Resources section above, glancing at the acronym 2U PERS WAR + will remind you to call for the following resources:

2nd Alarm

Utility Company

Police

EMS

Rapid Intervention Crew

Safety Officer

Water Supply Officer

Accountability Officer

Rehab

+ (any other resources the scenario calls for)

See how it works? Once you place your own format down on paper, you can go back and fill in any pertinent information that is specific to the scenario you are dealing with. If you have practiced enough, it will take you less than two minutes to write out your format. If it takes three minutes to read the scenario and you have ten minutes to prepare, you are now left with five minutes to "fill in the blanks" and complete your notes.

1. Read the scenario,

2. Write down your universal format

3. Insert incident specific information

Memorizing a format is important because it gives you an edge and helps ensure that you will not miss any key points during your oral presentation. Having it written down in an organized manner is priceless. If you miss something during your presentation, it will help you to go back and hit the point, even if it's out of order.

Rehearse your presentation

How do you rehearse? The best way is by video- or audio-taping yourself as you practice giving your presentation. I have found it helpful to ask a study partner or family member for help. You may find it difficult to watch or listen to yourself at first. After all, we are all our own worst critics. This step; however, is crucial because it will enable you to make adjustments and fine tune your presentation. Look for the subconscious body movements and facial expressions on the video tape. Listen for voice inflections and verbal flaws such as ums and ahs. Don't get upset if you aren't perfect – you won't be – just remember that your competition is doing something to prepare, you might as well do more.

Here is a simple daily ritual that will help you in the days leading up to your oral presentation:

1. Study: You can NEVER have enough knowledge.

2. Practice your acronyms and formats: Use mock scenarios, such as the ones in this book to practice writing your formats in the time frame you will be provided on exam day. You can do this anywhere (coffee shops, book stores, etc). At first, you will want to do this in an area where you can concentrate, but I advise you to also do this in areas where there are distractions. By doing so, you will be less likely to be intimidated or thrown off

if something happens on game day.

3. Practice your delivery: talk to a study partner, family member, video tape, audio tape... etc.

4. Make adjustments to improve your presentation.

5. Repeat steps 1, 2 and 3 DAILY from now until your presentation.

Game Day

So, today's the day! You've put in your time and whether you are ready or not, it's game day. Surely, you have tons of question … What do I wear? Do I shake hands with my assessors? Do I speak fast or slow? The tips below should help answer those questions and more.

Entering the room

First impressions are important. Although you will not be scored on your entrance into the assessment center, a professional entrance may create a good enough impression that may ultimately influence the assessors, resulting in a better overall score.

Although there has been much debate on how a candidate should enter the room where he/she is going to be assessed, there are a few simple things you absolutely should do in order to make a great impression. Keep in mind; this is essentially a job interview. The people you are competing against know you, but the assessors do not. This creates an equal playing field for all candidates. The senior member, who also happens to be best friends with the Chief may carry some weight around the firehouse; however, in the assessment center, that weight is reduced to a few seniority points that could be overcome by proper preparation from a well-prepared candidate.

Video camera: If your assessors are not in the room and you are giving your presentation to a video camera of some sort, don't be fooled into thinking you will not be judged on you overall appearance and entrance. Walk in with a smile and confidently take your place in front of the camera. Follow any directions you receive precisely and follow the tips below on body language.

SME's and assessors: Greet your Subject Matter Experts and assessor's professionally. Look them in the eyes and smile. If you are not perspiring, a handshake is okay, but use ONE HAND. Do not give the signature politician TWO HANDS and a fake smile. Superiority can be expressed in the initial handshake. Be firm, but not too tight. This is professional. (Palm down and too tight is superior, palm up and light is subordinate, either can work against you). Say hello before you sit. When you sit, make sure your back is upright and against chair. Plant your feet firmly on floor, do not cross your legs, and do not drum on the table or tap your feet. The key is to find comfort. If you are not comfortable, it will show and you will come across as nervous.

Once you sit. Do not adjust your tie, coats, etc… Once you are in position, be still, but not stiff. Square your shoulders, straighten your back, and make frequent eye contact, especially when being spoken to. Do not fidget in your chair. If there is a communications/body language expert in the room assessing your communication skills, be sure to look at them as often as you look at the SME.

Greeting your assessor

Do not try to engage in conversation with your SME. A simple "good morning" would do. Refrain from telling jokes. Trying to be funny while giving an oral presentation can be disastrous, unless you are a natural expert… and even if you are, this is not the time or the place.

Body language

Words are not the only way people communicate when they are speaking. Beyond the ums and ahs, there are gestures and expressions that tell stories. With very few exceptions, people consistently communicate their inner feelings quite openly. Each gesture is like a word in a language, and whether you are aware of it or not, we all subconsciously speak that language.

Consider this for a moment. Everyone has a distinctive walk that makes them easily recognizable to their friends. In the book titled How to read a person like a book, the authors speak about gesture clusters that enable people to read you. These clusters are a grouping of individual gestures that tell a story about you, and if you're not careful, they will cause your assessor to judge you regardless of the words that are coming out of your mouth.

Here are a few Examples:

Imagine a man who is sitting rigid, in an upright position with ankles locked. His eyebrows are raised and his hands are clenched together making one big fist, and he is rhythmically messaging one thumb against the other. Alone, these gestures are not very telling, together; however, they send the message that this man is NERVOUS.

Now imagine a woman who is sitting across the table from you who is unable to make eye contact. Instead, she is looking at the floor, her shoulders slumped. The woman is trying to tell you something, but she is clearly struggling to come up with words. It's obvious this woman LACKS CONFIDENCE.

In a third example, without saying one word, it's clear that the person who turns their back on you while you are talking and slams the door as they walk out of the room is ANGRY.

The point is simple. You must be aware of what your body is saying. You should also be aware of single gestures that send a strong message to others. Take for example, a person with arms folded high on their chest. This gesture is synonymous with stubbornness (think of a Baseball Umpire who is being yelled at by a manager). Let's look at several other types of nonverbal communication that are easily recognizable and often encountered:

- Frown: Displeasure or confusion

- Raised eyebrows: Fear, envy or disbelief

- Hand to cheek: Nervous, critical, pondering, or listening intently

- Inability to make eye contact: Hiding something

- Touching/rubbing nose: doubt

- Covering your mouth: Hiding something

- Pinching the bridge of your nose, eyes closed: self-conflict

- Clearing your throat (more than once): typically means anxious or apprehensive. This; however, is more than just a subconscious gesture. Mucus forms in the throat when a person becomes anxious or apprehensive. The natural thing to do when this occurs is clear your throat. The key is to prepare well enough so that you are 'less nervous'.

Gesture clusters, together, will tell you a lot about a person, but when it comes to single gestures, facial expressions tend to be the dominant body gesture. When giving an oral presentation during the testing cycle, consider the fact that most of your body will be hidden behind the table. If you are sitting upright, which you should be, your facial expressions will be the most obvious gestures.

Top sales people are well aware of the fact that a person's facial gestures are telling a story. When dealing with a prospect whose eyes are turned down and face is turned away, the sales person knows he is being shut out. The same sales person will recognize that the sale is virtually made when the prospects head is shifted to the same level as theirs and they are sharing an enthusiastic smile.

Quick reference speaking tips:

Here are 8 tips to help you deliver your message in a confident, professional manner.

1. Speak clearly and be natural.

2. Don't shout or whisper, make sure everyone in the room can hear you.

3. Don't rush your words or talk deliberately slowly. It's okay to pause at key points to emphasize the importance of a particular point you are making; However, be aware of your time constraints.

4. Don't try to be a flashy speaker. Changing your delivery (speed, pitch of voice, etc.) may work when giving a speech, but rarely does during an oral examination.

5. Keep hand movements to a minimum. Although it may be okay to sparingly use your hands to emphasize points, don't indulge in too much hand waving. This can be irritating and distracting. Your hands should remain on the table or by your side in a natural manner.

6. Make eye contact. You can look down at your notes, but look at your assessor(s) as much as possible. If there is more than one, don't fix on one individual. Eye contact projects confidence.

7. Avoid nervous body movement. Refer to the body language section of this article for tips.

8. Don't ramble. Make your point then move on to the next point(s). If you are asked questions or prodded at the end, your answers should be concise and to the point.

Presentation Aids

You will not be provided any presentation aids other than your notes, but they should be used for your reference only. Additionally, you may be allowed to have a pen and stop watch on hand. If so, place all three (your notes, pen and stop watch) in front of you in an organized manner. When you walk into the testing room, the assessor may hand you notes and ask if they are yours. Quickly scan through them and make a mental note of what pages you have (if more than one). Place your pen and stop watch on either side. If you are permitted to use a stop watch, it will be for you to keep time. When the assessor signals that your testing period has begun, start the timer. Occasionally glance down at your watch to keep track of time. (Note: this is the same procedure you want to follow when studying and practicing for your oral presentation. You don't want to fumble around, trying to find the start button at the beginning of your presentation. This will make you look unprepared, unorganized, and nervous. All of which will negatively affect your overall score.

Additional Tips

- Don't get frustrated. As you approach the day of your exam, you may find yourself becoming overwhelmed with the process of learning, retaining, organizing, and practicing your presentation. This is common. The worst thing you could do is allow yourself to become frustrated. While that may certainly be good for your competition, it is obviously not good for you. When you feel overwhelmed and frustrated, simply step away and take a break. Remind yourself that you are going the extra mile. The person that puts in the time and overcomes the hurdles he or she encounters deserves the rewards that come with the promotion. Be that person!

- Find a study partner/group. Studying with a group of firefighters from other departments who are also preparing for the same exam can provide you with a great advantage. You will be able to share information, bounce ideas off each other and provide valuable feedback and constructive criticism.

- Visualize. Whether you believe it or not, visualization is a technique used by some of the highest achievers in all areas of life. Think of an athlete, an MMA fighter for example. Leading up to a big fight, the fighter will visualize his hand being raised. If YOU can't see it in your mind's eye, how can you expect others to see it?

- Show up early. On the day of your exam, there are few feelings that could be worse than the thought of being rushed. Imagine being stuck in traffic, two miles away from the exam site, and you were supposed to be there five minutes ago. That kind of stress could easily be avoided by planning ahead. If the exam site is less than 45 minutes from your house, give yourself plenty of time to get there, and factor in Murphy's law (traffic, flat

tire, etc.) If your travel time is more than 45 minutes, consider staying at a hotel that is closer to the site.

- Enjoy the process!

Resources

There are many resources available to help you prepare for the promotional process and the moral requirement of becoming a better fire officer. Here are just a few:

www.FireOpsOnline.com

FireOpsOnline provides a wealth of free info available for dedicated firefighters. Check out the *Officer Development* page and you will understand why thousands of firefighters visit the page every month. Sign up for the newsletter and receive invitations to future Promotional Webinars and other valuable information.

In addition to your recommended reading list, here are two of my books that were specifically written to help people like you, who are serious about advancing their careers.

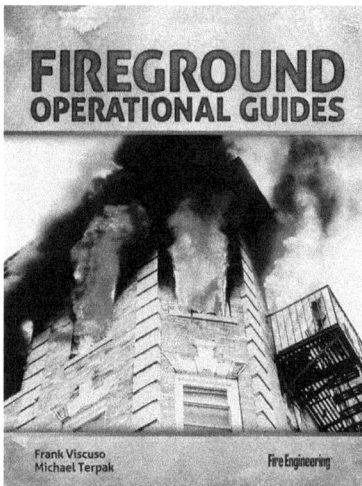

Fireground Operational Guides

By Frank Viscuso and Mike Terpak

This book features Operational Guides for 70 specific incidents. Each guide provides step-by-step actions you must take at all 70 incidents, which include 16 types of structure fires, natural gas emergencies electrical emergencies, water emergencies, carbon monoxide investigations, hazardous materials operations, structure collapse operations, chlorine emergencies, propane emergencies, bomb threats, and much more.

The book also comes with a CD that allows you to print out all 70 operational guides and a universal tactical worksheet that you can use in the field.

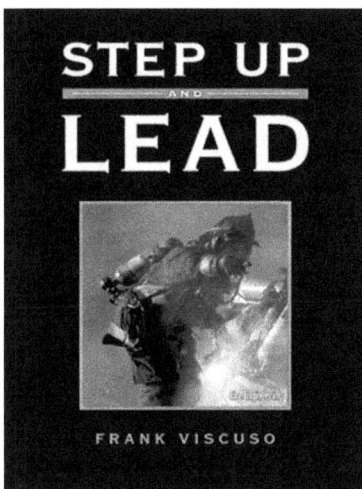

Step Up and Lead

By Frank Viscuso

Industry best-seller 'Step Up and Lead' is loaded with information to help individuals prepare to move into leadership roles. This book provides a chapter dedicated solely to Leadership Skills. In that chapter, you will find detailed information on taking promotional exams, dealing with subordinate problems, and tackling administrative tasks. Bobby Halton, Editor-in-Chief of Fire Engineering magazine says, "Here is an outstanding book written by a world class firefighter about HOW to lead! This is a must read for every firefighter."

For more information, visit www.FireOpsOnline.com.

www.ingramcontent.com/pod-product-compliance
Lightning Source LLC
Chambersburg PA
CBHW061812210326
41599CB00034B/6969